Handbook on the International Exchange of Publications

Handbook on the International Exchange of Publications

5th completely new edition

Edited on behalf of IFLA

by
Kirsti Ekonen, Päivi Paloposki
and Pentti Vattulainen

K · G · Saur München 2006

Bibliographic information published by Die Deutsche Bibliothek
Die Deutsche Bibliothek lists this publication in the Deutsche Nationalbibliografie; detailed bibliographic data is available in the Internet at
http://dnb.ddb.de.

∞

Printed on permanent paper
The paper used in this publication meets the minimum requirements of
American National Standard – Permanence of Paper
for Publications and Documents in Libraries and Archives
ANSI/NISO Z39.48-1992 (R1997)

Printed in Germany

Printed / Bound by Strauss GmbH, Mörlenbach

ISBN-13: 978-3-598-11752-7
ISBN-10: 3-598-11752-3

TABLE OF CONTENTS

Part Two: History and Examples of Our Days

Part Three: Directory

PREFACE

The first edition of the Handbook on the International Exchange of Publications was published in 1950 but it was only the second edition in 1956 that constituted the Handbook in the real sense, enriched by a practical guide and a significant directory. Since then there have been two updates, one in 1964 and another in 1978 (UNESCO). There has not been a fifth edition since.

IFLA Acquisition and Collection Development Section discussed about the need to update Handbook on the International Exchange of Publications in 1990ies. The Section estimated that exchanges continued to be a mode of collection building, which still continued to be practised by almost all major libraries up to third millennium. However during the last 25 years there have been drastic technological progress and very important changes in the politico-economic arena, all of great influence on the object and methods of the international exchange of publications. Hence a new handbook is necessary.

The structure of the Handbook has not changed much compared with the previous editions. The first part of the Handbook consists of updated text of the 4th edition's practical guide. The text was revised mainly by library director Anneli Virtanen, who has been a member of the Standing Committee of the Section. Parts of the text are revised by the staff of the Exchange Centre for Scientific Literature in Finland Inga Kontula and Sari Lehtinen and the head of National Repository Library of Finland Pentti Vattulainen. Part Two - History and examples of our days - consists of articles which originate mostly from the IFLA Acquisition and Collection Development Section's workshop in 2001 IFLA Boston conference. Part Three – Directory is an update of the exchange centers in the world. It is by no means an extensive list but hopefully it gives an overview on the ways how exchange activities are organized in various countries. There is also contact information for those that need further information on a specific case.

As in the previous edition this edition is intended for the use of (a) librarians who want to set up an exchange centre, especially in developing countries; (b) exchange librarians who want to enlarge their knowledge; (c) students of library schools; (d) officers responsible for library work; (e) information officers and documentalists.

For various reasons the time between the editing process and publishing became long and hence some of the references may refer to older editions of certain publications. In this kind of publication where every detail is important there may be errors and omissions. We would welcome any corrections which readers may wish to point out.

I express thanks to the members of the planning team that sketched the contents of the new edition of the Handbook. The team included prominent librarians, IFLA veterans and the librarian of the Exchange Centre for Scientific Literature of Finland: Eeva-Maija Tammekann, Pia Sodergard, Johanna Lilja, Jorma Hirsivuori, Maire Aho, and Jarl Pousar.

Special thanks for Päivi Paloposki, MA and Kirsti Ekonen, phil.lic, who have collected the material and filled caps for the directory as well as participated in many ways to getting the Handbook finally done. Ms Ekonen's contribution has been exceptionally valuable.

The new version of the Handbook is finally ready even though there have been many drawbacks during the preparation. The librarian on the Exchange Centre Jarl Pousar passed away in April 2004 during the editorial time. His experience and vision of the exchange activities around the world was very important for the work.

The previous editions of the Handbook were published by UNESCO. As UNESCO is not any more involved with exchange activities, IFLA - through its Acquisition and Collection Development Section - should take care of some of the items connected with exchange activities. We hope that this book is useful for libraries around the world even though there is a strong Finnish bias in the contents.

Kuopio January 2006

Pentti Vattulainen

Chair, IFLA Acquisition and Collection Development Section

PART ONE: PRACTISES

BACKGROUND

The international exchange of publications is a form of international scientific and cultural co-operation. It promotes the free flow of ideas and scientific information among institutes belonging to different nations in virtue of a formal contract or a free arrangement whereby the parties concerned give one another publications in printed form or reproduced in some other way. This contract is based solely on the mutual consent of the parties; it is not, in principle, subject to any formula and may be embodied in an informal letter; what is important is that both parties should execute the agreement.

This handbook concerns the exchange of publications mainly for scientific purposes, using the term 'scientific' in a wide sense (including the humanities and social sciences). Exchange from commercial motives finds no room here. The printed matter for exchange may comprise monographical or periodical publications; it may emanate from official or private sources, through the book trade or independently of it.

The question may arise: Why should such matter be sought by way of exchange, when it can be procured by purchase? Would it not be easier for the institutions concerned to obtain the material required through normal trade channels? Is not a book, in the first place, an object of commercial exchange and the book trade, therefore, the legitimate intermediary between producer and consumer? Has not barter everywhere been replaced by transactions for money? Why, especially in the case of books, should the primitive form of transfer be preserved? This handbook is meant to be the practical answer to these questions.

The exchange of publications constitutes one of the ways in which science manifests its universality. It would be a misconception to regard it as merely of economic significance. It is a two-sided transaction in which the making of a profit does not arise; it is half-way between a sale and a present. Its pattern is that of two partners of similar standing 'getting together'. Its necessity may be justified both from the receiver's and the giver's point of view.

On the receiver's part it might be argued that there are certain categories of books which are unobtainable in the trade at all; while there are other categories which the trade would have difficulties in getting hold of. There are certain situations in which purchase - at least for some time - is rendered impracticable (for example, a library may lack ready funds, but it may have at its disposal a number of copies of its own publications or duplicates, which might be made available; or a library may find the purchase of foreign books temporarily impossible owing to the lack or scarcity of foreign currency). In such cases, exchange is, if not the only means, one of the few available means, of obtaining books.

From the giver's point of view it is possible to get even closer to the idea underlying the exchange of publications. He who issues his own publications wishes to see them in the hands of fellow 'scientists engaged in the same field, without the intervention of the book trade which, in this case, hardly stands to profit much. Or else, the owner of printed matter that has become redundant wants to find a partner having something to offer, with a view to effecting an exchange, without financial benefit to either.

We ought to accustom ourselves to regard the exchange of books as akin to the exchange of personalities, which latter it partly replaces and partly supplements. Both serve for the exchange of experiences and, incidentally, for promoting mutual understanding.

In the following sections we shall deal with the actual working of exchange. But before doing so, here are a few maxims valid for exchange in general:

1. Look upon the exchange of publications as a form of scientific conversation with your equals.

2. Do not send superfluous material and/or material of secondary importance in exchange.

3. The exchange material should meet a need of the exchange partner.

4. A generous attitude should always be adopted towards the international exchange of publications. However, some kind of balance is desirable, in order to preserve the distinction between exchange - implying reciprocity - and mere gift.

5. Do not aim at having as many exchange connections as possible; but rather endeavour to find the partners who are best suited to your particular purpose.

6. A law, valid for all conceivable cases of exchange, does not exist. Do not attempt to mould your entire exchange activities on a uniform pattern. Rationalize, by all means, the pertinent technical operations, but do not fail to regard each one of your partners in exchange as an individual case.

SOURCES OF DOCUMENTS

Ed. by Anneli Virtanen

1. Types of Exchange

International exchange between libraries has diminished during last decades, and its future has been even made questionable. First of all exchange has been seen as a not cost-effective way to acquire material due to the great amount of work it needs. Increasing electronic publishing and easy access to the Internet has also made exchange less important. Nevertheless exchange will play a certain, although reduced, role in the overall acquisitions process. To some libraries exchange is a means to get publications issued in small printings, regional publications or grey literature. Libraries, which have hard currency problems, can add their acquisition funds with exchange. It is good to remember that an exchange is not only a financial matter, but also cultural co-operation between libraries or generosity towards institutions especially in developing countries. Exchange is also a means to get own publications spread to other libraries and institutions all over the world.

Exchange partners are considered generally to be libraries and institutions with which the mutual supply of materials has been arranged. In principle every type of library can use almost every object of exchange, but usually a certain specialisation takes place.

When establishing exchange connections, i.e. the selection of appropriate partners and materials, the reference works[1], websites[2] and bibliographies are of great help. For the former, *World Guide to Libraries, The World of Learning,* and similar works can be recommended; in the Internet there can be found for example *The National Library Catalogue Worldwide, Libweb* on the Berkeley Digital Library SunSite with more than 6 600 web-sites in 115 countries, and Sauers' *The WWW Library Directory* more than 8 800 libraries and library-related web-sites in 130 countries. For the latter comprehensive national bibliographies, libraries' and publishers' catalogues, and lists of serial publications, official publications and learned organizations are available both in printed and digital forms.

---

[1] World Guide to Libraries / comp. by Willemina van der Meer. 20th ed. München: K.G. Saur, 2006, and CD-ROM edition World Guide to Libraries Plus. Annual update; The World of Learning 2004. London: Europa Publications, 2003; World Guide to Special Libraries. 7th ed. München: K.G. Saur, 2005

[2] http://www.library.uq.edu.au/ssah/jeast/; http://sunsite.berkeley.edu/Libweb; http://www.webpan.com/msauers/libdir/; http://www.bundestag.de/bibliothek/library/index.html

As concerns the quality and quantity of the material exchanged, efforts are made by the majority of libraries to set up a balanced account, although it is unlikely that an exact balance can ever be obtained. For this there are three ways:

a) Piece-for-piece exchange, i.e. book for book and pamphlet for pamphlet, with which the subscription-for-subscription exchange of serials or journals can be grouped (title for title, annual volume for annual volume, if necessary also number for number, or one journal appearing at frequent intervals for two journals only a few numbers a year each).

b) Priced exchange, in which each partner agrees to supply the other with publications of a set monetary value, on the basis of a certain ratio of the currencies involved, within a stated period of time. This method of compensation is practical especially cases where much heterogeneous material, for example, books, journals, copies, etc. is exchanged. Occasionally the ratio may differ from the official rate of exchange.

Priced exchanges, however, offer some difficulties. While books in a number of countries become more and more expensive, their prices are kept low in some other countries, so that the discrepancy increases constantly. In this sense, priced exchanges can to some degree be problematic. This difficulty can be avoided by fixing individual ratios, not based on official rates or prices, among the partners.

Despite of some difficulties in ratios, this type of exchange is increasing and takes place of the other types.

c) Page-for-page exchange, i.e. one page for one page, with possible variations, if special objects are involved, for example, one table for two pages, two or three pages of a new book for one page on an old one, etc. This method of compensation is appropriate in cases where book prices differ very much between two countries.

In addition, there is the 'open' exchange in which there is little or no accounting. The philosophy of such an arrangement is that if each partner supplies the other with one copy of all its publications or certain specially designated ones, the exchange over a period of time will tend to be balanced, anyhow. Although an exchange should be founded on a basis of reciprocity, generosity is advisable in certain cases, above all towards institutions in developing countries . With regard to the bibliographical tools, the choice of appropriate partners, and the establishment of a balance it will suffice to supplement to above explanations merely with additional remarks, which apply only to individual objects of exchange and/or partners.

2. Academic Publications

By academic publications are meant here all publications appearing at the instance or under the auspices of academies, research institutes, learned societies, and universities.

a) Academies, Research Institutes and Learned Societies

Academic publications are, in a narrow sense, publications issued by an academy directly, for example treatises, proceedings of congresses, etc., yearbooks, monthly reports, communications, statutes, chronicles, memorial volumes, publications about the internal life of an academy or one of its institutes, and general series of miscellanies. In a broad sense they are scientific publications issued by an academy or a scientific institute in the fields of the humanities, technology, natural sciences, etc., for example books, journals, series, transactions as well as reports on scientific research and other scientific events. Many libraries, archives and museums are engaged in research work, for example the Biblioteca Vaticana and the British Library.

Learned societies publish analogous literature in their respective fields. Their publications have come more often out by commercial publishers.

In the selection of partners for the academic publications, an academy or a learned society[3] must be guided by the specific requirements of its clientele in general and its members in particular. In some cases they entrust a neighbouring research library in the locality or a national centre with the exchange. Partners and bodies responsible for the exchange of publications of academies, research institutes and learned societies in this broad sense are, therefore, the general scientific libraries - the national and State, university and municipal libraries, in addition to the libraries of academies. In many cases learned societies have no library of their own. Frequently, however, some of their publications such as journals and transactions are obtainable only by membership of the society or organization. The principle of a generous attitude and mutual respect of the partners, of mutual giving and receiving, should be applied in this category of exchange.

If a certain balance is required, the subscription-for-subscription method is to be recommended, as most academic writings are published in series or journals. Institutions concerned with special fields of science, for example, special academies, which as a rule do not issue more than one or two publications in a year, usually reckon on this basis.

b) Universities

By publications of universities are meant all publications appearing at the instance or under the auspices of universities. In some countries, such as the United States of America, Sweden and Switzerland, publications written by members of a university are also numbered among university publications. In a narrow sense they are such publications as are issued directly by a university or college, or by one of its independent units, for examples theses, inaugural dissertations, lectures by professors, programmes, calendars, study guides, yearbooks, com-

---

[3] World Guide to Scientific Associations and Learned Societies / ed. by Helmut Opitz. 10th ed. München: K.G. Saur, 2005

munications, memorial volumes, and general series of miscellanies. In a broad sense they are scientific publications issued by a university or college or a scientific institute attached to it.

The bulk of university publications is made up of dissertations and theses. They can be typewritten, and may therefore be difficult to obtain, printed or/and in electronic form. Today ever-growing amount of dissertations is published in electronic form, and they can be read on the websites of universities. Dissertations are reproduced also on microfilms or on microcards, and as CD-ROM versions. Moreover, dissertation abstracts are printed and published by a number of universities in the form of yearbooks, which in their turn are offered for exchange. Finally, there is also a tendency to publish dissertations as articles in journals or series and in certain cases use reprints for exchange.

Special regulations are required for dissertations published by some university presses, which provide for a certain number of copies to be placed at the disposal of the respective libraries for exchange, sometimes at considerable discounts. In this connection an agreement may be concluded with the university press to distribute publications only after a certain time has elapsed, generally two or three years after the date of the publication.

Exchange partners are in most instance university libraries, which frequently carry out exchanges for the entire university centrally, i.e. inclusive of the publications of the other units such as faculties, institutes, seminars, departments and experiment stations. In many cases, however, the latter institutions handle their exchanges directly usually through their own libraries, if such exist. Occasionally also larger libraries (e.g. a State library) can appear as partners, if university publications obtainable through the book trade are involved. There are different ways of handling theses. In most cases, a separate division is established for their administration and exchange.

Where there is a separate centre for the exchange of dissertations or where the national exchange centre is responsible for it, the choice of partners will be more rigidly fixed and incoming material will be distributed by certain rules.

The above-mentioned diversity in the forms of publications makes it almost impossible to find a balance of value applicable to the exchange of dissertations. The partners can try to balance their exchange with other materials. For all publications of universities, the same generosity as with academy publications is considered appropriate. Sometimes the piece-for-piece or the subscription-for subscription method may be applied to university publications in a broad sense.

In many cases the publications of universities are listed with the other material in the general editions of national bibliographies, or in a separate edition, on the web-sites of universities, occasionally also in the lists of official publications (e.g. in Denmark). Frequently, however, they are registered in special lists, which are issued periodically, often annually. Certain lists and catalogues give summaries of unprinted dissertations such as *Dissertation Abstracts International* (http:/

/www.asu.edu./lib/resources/db/dissabs.htm) in the United States. Useful surveys of national and international publications, which list current theses and dissertations, are contained in the manuals of bibliographies

3. Commercially Issued Publications

In many cases libraries are allowed to use part of their acquisition budget or a special budget for the purchase of book trade literature for exchange purposes. In practice a library buys such publications as are requested by the partners, or they are significant to libraries and, according to experience, are often asked for. National and State libraries can also use their free examples for exchange.

Among these may also be numbered the publications of other categories (publications of academies and learned societies, of universities, research institutes, corporate bodies, official bodies, and international organizations) as far as they are distributed by the book trade. In certain instances, such publications need not be purchased, i.e. if academies, universities, research institutes, etc., donate a certain number of copies to their respective libraries or to some library or if libraries, according to law, receive more than one copy of the national output. There are many ways to find commercially issued publications. In addition to what is mentioned earlier, the bookstores in the Internet[4], for example, are useful.

This type of exchange material is not limited to a specific kind of library. Therefore, all kinds of libraries can be suitable partners. When books and journals are bought through commercial channels, a certain balance will have to be reached, e.g. by applying the piece-for-piece, subscription-for-subscription, or page-for-page methods. In the last case, a fictitious price may be established for one page, or, if a book is very expensive, the number of pages counted may be increased. When money is used to deliver material, priced exchange can be recommended.

4. Official Publications

Official publications are highly important as a documentary source material for researchers in many fields, but their chief importance, however, lies in the fields of social sciences and administration.

The ruling political powers of a country (governments, parliaments) need the publications of other countries for their daily tasks and are, on the other hand, interested in distributing their own material. Therefore, the exchange of official publications is organized officially and centrally in most countries. A central organization is necessary to obtain and dispatch exchange material, especially in

---

[4] For example http://www.amazon.com; http://harvard.com; http://www.chapitre.com

countries where the publication of official documents is not centralized. Similarly, a central deposit can be designated to receive foreign documents if it is wished to avoid their dispersion among too many recipients. The Unesco Convention Concerning the Exchange of Official Publications and Government Documents between States of 1958 gives no more than a description of what has to be considered as 'official publications and government documents when they are executed by the order and at the expense of any national governmental authority'. However, the term 'official publication' varies in different counties and that is considered even in the Unesco Convention by adding 'and other publications as may be agreed'. So the decision is left to each individual exchange partner, and it is up to the Contracting States to determine the official publications and government documents that shall constitute exchange material.

For international exchange purposes, the following are to be regarded as official: the publications of central, federal and regional governmental bodies and those of their subordinate institutions; publications of public law corporations, institutions, foundations and companies. Not discussed here are the so-called semi-official publications, i.e. publications influenced by one of the bodies or institutions ('on behalf of', 'at the instance of', 'in connection with', etc.), or publications of offices which are not official bodies.

As a rule, but by no means in every case, at least part of official publications is issued outside the book trade and is therefore difficult to acquire through regular commercial channels. On the other hand, in some countries almost all official material is published by a government printer, H.M. Stationery Office in the United Kingdom being the oldest case in point.

Official publications and government documents either serve an administrative aim, for example, the promulgation and explanation of laws, or an informative purpose, for example, statistics. Purely scientific literature is not included in the exchange of official publications, but form part of the material exchanged directly between respective institutions of this part.

Administrative publications are mainly: parliamentary documents, laws and official gazettes, bodies of law, decrees and notifications, decisions of the Court of Justice, treaties, constitutions and statutes and other legislative and regulative documents. Informative publications include material such as reports, statistics, annals, news bulletins, state handbooks, yearbooks, white books, national bibliographies, catalogues, maps and plans, pamphlets and other material as may be agreed. These kinds of publications can be produced, except by independent States, also by organizations like European Union. The difference between official publications and issues of international official bodies is not always clear. Usually also the voluminous and important group of patents and standards comes under this category of publications.

The convention concerning the exchange of official publications is an agreement between two States. Therefore, the choice of participants - i.e. of the countries with which a system of exchanges is to be introduced on the basis of the agreement - is influenced by governmental policy and in many cases is, indeed, made by the government itself. In many countries parliamentary libraries are exchanging official publications.

From the bibliographic point of view, official publications are not always easy to trace. In this viewpoint they can be partly handled as grey literature.

In the majority of cases, a non-commercial approach prevails in the exchange of official publications, because their quantity differs from country to country, and libraries require all relevant publications. The 1958 convention encourages the conclusion of agreements on selective and partial exchanges, but also complete sets are dispatched. Therefore, the volume is usually fixed at the beginning. If a larger country insists on a balance, then the smaller country can fill the gap by sending other types of material, for example, scientific literature. The best and quickest method for establishing a balance will be to count approximately the number of the volumes or series received and sent, and at most to exchange one series against other. But the more advanced States should expect to give more than they receive.

5. Publications of International Organisations

Organizations are considered international if their membership reaches beyond national boundaries. One way to classify organizations is to examine what purpose they have. May have aims and activities that encourage co-operative behaviour, such as the Nordic Council or they may reduce conflicts as the United Nations. The most usual way to classify organizations is to divide them into international governmental organizations (IGOs) (http://www.gksoft.com/govt/en/multi.html) and non-governmental organizations (NGOs) (http://www.ngo.org/). IGOs are created by intergovernmental agreements and their members are governments.[5]

Most of the IGOs issue their own publications. A unique case is the United Nations. The United Nations organization system consists of main bodies, spezialized agencies, programmes and funds. Each agency issues publications and documents, which represent their activities. Many of these publications are unavailable from any other source. The publications of the United Nations contain mimeographed documents, official records, sales publications, treaty series and periodicals.

---

[5] The European Directory of International Organizations 2003. 5th ed. London: Europa Publicatiobs, 2003; The Information Systems of International Inter-Governmental Organizations: A Reference Guide. Stanford, CT: alex Publishing, 1998

See also http://www.libsci.sc.edu/bob/IGOs.htm; http://www.stanford.edu/group/Jonsson/igourl.htm;http://docs.lib.duke.edu/igo/guides/ngo

The United Nations specialized agencies, the other IGOs and the NGOs publish a great number of bibliographies, catalogues and lists of publications, general and special indexes to documents, publications which explain the purposes of the organization and describe its activities (e.g. reports, research programmes, leaflets), basic instruments (e.g. official bulletins, rules and regulations, manuals, constitutions), official records (e.g. proceedings, minutes, resolutions and decisions, seasonal documents), journals bulletins, transactions, yearbooks, handbooks, statistics, monographs and similar kinds of material. The publications of the United Nations and its spezialized agencies are included in the *United Nations Documents Index* (UNDEX)[6] and the *Guide to the Archives of International Organizations: Part 1. The United Nations System*[7].

The UNBISnet-database is a free electronic publication of the Dag Hammarskjöld Library http://unbisnet.un.org/), which contains bibliographic records, voting records of the General Assembly and the Security Council and index to speeches made in main bodies of the United Nations. Most international organizations publish bibliographies at present also in the Internet. E.g. Unesco (http://unesdoc.unesco.org/ulis/index.html), International Labour Organization (ILO) (http://www.ilo.org/public/english/support/publ/intro/index.htm), World Trade Organization (WTO) (http/w//ww.wto.org/english/docs_e.htm), and Council of Europe (EC) (http://book.coe.int/EN/index.php). Their web pages contain documents about decisions of official bodies, conferences and a great deal of other material.

The exchange procedure is usually the same for NGOs as that for scientific and other organizations. For IGOs exchanges may be more difficult to arrange. However, after an agreement has been made, exchange is conducted in the same manner as with NGOs, scientific and other societies.

National associations, institutes and libraries are all interested in exchange with international organizations but there may be difficulties to build up exchange relations. Unesco and the Union of International Organizations have published helpful guides, which contain names of international organizations in different languages, abbreviations of the names and addresses[8]. The first difficulty is to

---

[6] United Nations Document Index: Documents and Publications, Author Index, Title Index, Subject Index / United Nations, Dag Hammarskjöld Library. Vol. 1.2. New York, NY: United Nations, 1998-

[7] Guide to Archives of International Organizations. Part 1: The United Nations System. Paris: Unesco, 1984

[8] Yearbook of International Organizations: Guide to Global Civil Society Networks / ed. by Union of International Associations. Vol. 1-4. 42th ed. München: K.G. Saur, 2005; The Europa World Year Book. Vol. 1-2. London: Europa Publications, 2003; Europahandbuch. Köln: Heymann, 1998; Introduction to International Organizations / ed, by Lyonette Louis-Jacques and Jeanne S. Korman. New York, NY: Oceana Publications, 1996; Schiavone Giuseppe, International Organizations: A Dictionary. 4th ed. London: Macmillan, 1997; Transnational Associations. Brussels: Union of International Associations, 1997

locate the name, often there are two or more different languages, and the address is not always the same. The titles of publications may be multilingual and the place of publications may change. Besides printed guides information of the organizations and their publications can be found in the Internet (http://www.uia.org).

International organizations distribute their publications in the following ways: sales through book trade - in each member state often one bookstore acts as the exclusive sales agency, distribution to members in return for the membership fee, distributions to depository libraries which automatically receive all publications, and exchange. The depository libraries get documents in printed form but new documents are usually also issued in the Internet. Older documents have not been digitalized retrospectively backwards yet.

International organizations have depository libraries in the member states. One country may have several depositary libraries in different countries. The library chosen as depositary must be a national library, another major research library or a parliamentary library open to the public. Depositary libraries do not get necessarily all publications free of charge. The organizations have many different ways to distribute their publications.[9]

Electronic publishing both in CD-ROMs and through the world-wide-web has changed the distribution and exchange of publications. The publications of international organizations usually are available free of charge in the Internet; in some cases publications are liable for charge. All printed publications and documents are not published in the Internet, e.g. sales publications of international organizations are issued only in paper form.

A good example of electronic publishing is the United Nations www-pages (http://www.un.org/documents), which contain seasonal documents of the General Assembly 52.-58. sessions, verbatim records 55.-58. sessions and resolution and decisions since 1946, the resolutions of the Security Council since 1946, the resolutions of the Economic and Social Council since 1982 and a selection of decisions of the International Court of Justice since 1947.

The United Nations Optical Disc System (ODS) is a system for storing and retrieving the United Nations documents. The database compromises the full text of all the United Nations parliamentary documents since 1992. The resolutions database contains the final official record version of all resolutions and decisions adopted by the General Assembly, the Economic and Social Council and the Trusteeship Council since 1946. The documents and official records are stored in

---

[9] Instructions for Depository Libraries Receiving United Nations Material (ST/LIB/13/Rev.1); Principles Governing United Nations Depository Libraries (ST/A1/189/Add.11/Rev.2, 18 August 1995)

six official languages of the United Nations. The users intending to use ODS must have access through a provider, but a part of the material of ODS is possible to use free through the UNBISnet-database[10].

6. Grey Literature

By the Luxembourg Convention on GL adopted by the Third Conference on Grey Literature, November 1997, "The grey literature embraces governmental, academic, business, and industrial documents, both in printed and electronic forms, remaining beyond the bibliographic control not available through the regular book-selling network. The grey literature comes within the large diversity of information materials such as: political documents, statements, research reports, statistical data specifications, academic theses and dissertations, technical specifications and conference proceedings."[11] This kind of literature is often meant for experts to inform rapidly a specific scientific community. Grey publications may contain comprehensive up-to-date information on research. The number of copies, if printed, is definitely limited. On the other hand many publications, e.g. technical and commercial documentations can be printed in thousands of copies. It is not easy to delimit precisely the boundaries of grey literature and many publications introduced in the other groups can considered to belong to the category of grey literature. Some publications of libraries, archives and museums can belong to grey literature, e.g. list of accessions, checklists, exhibition catalogues, etc. It is literature, which is included in bibliographies variously. In order to limit the difficulties in retrieving this kind of literature GL bibliographies and databases have been founded. In the Internet there are both portals and specialized websites of different fields.

One of the most important sources of the grey literature is The Information System on Grey Literature in Europe (SIGLE), which was founded in 1980 to collect and disseminate information on GL. It is a bibliographic database covering European non-conventional literature in the fields of pure and applied natural sciences and technology, economics, social sciences and humanities. It is accessible on the web-site and SilverPlatter delivers it on CD-ROM. The database is produced by European Association for Grey Literature Exploitation (EAGLE) (http://www.kb.nl/infolev/eagle/frames.htm). There are also printed bibliographies on

---

[10] Simultaneous availability of parliamentary documentation in electronic form in the six official languages on the United Nations' web-site / United Nations web-site / United Nations, General Assembly (A/C.5/56/12)

[11] Perspectives on design and transfer of scientific and technical information: Proceedings of the 3. International Conference on Grey Literature, Luxembourg, 13-14 November 1997. Amsterdam: TransAtlantic, 1998

the grey literature, for example *The Annotated Bibliography on the Topic of Grey Literature*[12]. The Trans-Atlantic Grey Literature Network Service (GreyNet) (http://www.greynet.org/pages/1/index.htm) rises from the co-operation between European and American professionals on GL. GreyNet includes i.e. a selection of web-based resources in grey literature. A significant part of the most traditional types of grey literature is published in the Internet as full-texts or at least abstracts of it. The search on GL documents is often oriented towards the issuing bodies and not the author.[13]

Grey literature consists very heterogeneous material, both in printed and electronic forms and there are many kinds of possible exchange partners, but scientific libraries are the most likely ones.

7. Other Exchange Materials

a) Duplicates and Surplus Material

Duplicates are second, or subsequent, copies of books already in stock, identical in edition and involuntarily acquired. They are mainly acquired through unsolicited gifts, by the global acceptance of special collections, literary bequests, etc., by second-hand purchases in bulk as well as by occasional mistakes in the acquisitions department. The term 'duplicates', as used here, excludes purposely acquired multiple copies of important works for reference departments, etc., as well as antiquated and unusable material (e.g. old schoolbooks, textbooks, editions of classics), which should be eliminated.

Apart from duplicates, surplus material consists of literature that has been withdrawn from the library collections because it is no longer needed and publications that come into the possession of a library but are not added to its collections because they fall outside its scope. National libraries or libraries performing similar functions may have still another means of acquiring surplus material: copies originating from legal deposit, i.e. copies of the entire national output supplied by publishers and/or printers, free of charge, according to law.

The exchange of duplicates and surplus material is an indispensable method of acquisition for all libraries which have either suffered losses or wish to fill gaps in their periodicals holdings, take over new subject fields, have missed certain

---

[12] Annotated bibliography on the topic of grey literature / comp. by Dominic J. Farance. 4th ed. Grey Net, 2000; Information Sources in Grey Literature / by Peter Auger. 4th ed. München: K.G. Saur, 1998

[13] Alberani, Vilma & De Castro, Paola, Grey Literature from the York seminar (UK) of 1978 to the Year 2000: http://www.ifla.org/VII/d2/inspel/01-4alvi.pdf

acquisitions in the past. It is a particularly useful method for newly created libraries, especially in developing countries, since most of them almost completely lack scientific literature published in the past. As an active means of acquisition it has not very important role.

Redistribution schemes exist, for example the United States Book Exchange. In some other countries similar institutions are also responsible for a centralized international exchange of duplicates, for example some national exchange centres. For the selection of exchange partners, the national exchange centres usually possess comprehensive lists of addresses of libraries in other countries interested in an exchange of duplicates and other surplus material, and can answer questions from abroad from their general knowledge of the library situation in their respective countries.

A priced exchange should normally be limited to the most precious items - complete volumes of periodicals, rare old books, etc. In such cases a price corresponding to the intrinsic value must be agreed upon. If applied to average books, it is useful to fix a set price, based on size for monographs and frequency in the case of periodicals, or to make use of the page-for-page method.

Exchange can be greatly assisted if generous estimates are allowed of the value of the material received. In this sense it is customary to exchange volume against volume. As a library deals with its equals, it has to rely on the fairness of the other party.

b) Electronic Publications

Ed. by Sari Lehtinen

Electronic publishing and distributing via Internet is widening. It has effects on obtaining publications and in exchange activities.

Electronic publications (mostly journals) issued by the big commercial publishers demand expensive access rights. Libraries have formed large national or international consortiums, which buy the licences for an extensive use in all libraries of the consortium. It is almost impossible to make an agreement for accessing a commercially issued electronic version in the frame of exchange. The exchange partner, the publisher and the distributor are often three different institutions, of which the two latter ones are interested in receiving also some profit. In the case of a hybrid version, some publishers give a free access to the electronic version for subscribers of the paper version. In most cases this matter does not concern those subscriptions acquired on exchange.

There are some possibilities to have free access to those commercial electronic publications which are issued and delivered by the exchange partner self. In these cases the exchange can be based on a print publication for an electronic version as well as an electronic version for an electronic version.

An increasing number of journals and other publications, which have a free access, can be found in the Web. Many institutions and universities publish free full text material in the net. There are also some materials only in electronic form. This may cause problems in obtaining the material in many countries.

Among the scientists and libraries the demand for free access to scientific research articles and other information available on the Internet has been the motive to establish the so-called open access movement. Open access (OA) publications are scientific refereed journals, collections of articles and other reports and papers of institutions. More information is available e.g.at the Budapest Open Access Initiative (http://www.soros.org/openaccess/). Directories for open access publications are for instance the DOAJ – Directory of open access journals (http://www.doaj.org/), which covers free, full text, quality controlled scientific and scholarly journals in all fields and Unesco Social and Human Science Online Periodicals (http://www.unesco.org/shs/shsdc/journals/shsjournals.html), which provides free access to full text of hundreds of scientific periodicals. BioMed Central (http://www.biomedcentral.com/) maintains a directory for open access and other biological and medical journals. BioMed Central has also a complete publishing system for open access journals and articles and some other services for authors (per review, archiving etc.). PloS Public Library of Science (http://www.plos.org/) publishes the open access journals PloS Biology and PloS Medicine into which scientists or institutions can insert their articles. Both BioMed Central and PloS have sponsors and they charge the authors of articles or their organisations and the journals.

Open access publishing and directories give an economic and easy way for readers and libraries to access publications in the Web. The lack of adequate it-technology is still a problem in many countries. And in the end, can we any more talk about exchange of publications, when it comes to the open access publishing? The libraries have only to take a charge in directories and links to the publications in the Web.

c) Other Materials

This category compromises rather heterogeneous material. It can contain reprographic material in different forms and other non-book material.

Reprographic Materials

Any copyable publication or manuscript can be acquired in xerocopies, microform or in electronic form in accordance with the rules of copyright in each country. Therefore, reprographic material can refer to any of the types of material already mentioned. Libraries, which lack the originals, are naturally interested in obtaining the corresponding copies. Some of the larger producers have developed microfilm projects which enable new libraries or libraries to develope retrospective research collections by acquiring large quantities of publications in microform or as CD-ROMs.

Non-book Materials

These are means of communication, audible and/or visible, which do not come within the definition of a book or serial and which require special handling. They either appear as independent objects or are, frequently annexed to a printed work. The significance of audio-visual aids as illustrative material in research and teaching is constantly increasing. These aids can be grouped as follows:

1. Graphic art, frequently published as broadsheets, i.e. pictures as art prints (frequently in picture files), posters, picture postcards; and printed music, maps and atlases are collected in the respective special libraries as well as in the music and map departments of the large academic libraries. Therefore, they, too, can be objects of exchange.

2. Visual materials, frequently published in series: slides, photographs, films.

3. Sound records: records, tapes.

4. Multi-media materials, i.e. multi-media kits for learning languages; programmed learning aids, charts, specimens, and models; television cassettes, sound/slides series.

All above-mentioned materials can be issued also in CD-ROM and other electronic form.

From the bibliographic point of view, the material can be difficult to trace and obtain. However, there are bibliographic tools for microforms[14], audio-visual materials and CD-ROMs. Records and maps are generally registered either in the national or in special bibliographies, the same applies to printed music. It may be difficult to exchange this kind of material with libraries. An exchange can be attempted with the respective institutions or organizations.

In exchanging these materials, a balance will generally be established on the piece-for-piece basis without any difficulties, for example, film-strip for film-strip, record for record, map for map, or photograph for photograph. If the items differ too much in size, either a ratio of one to two or more copies, or a priced exchange is recommendable. In many cases, however, a rough estimate of the material to be exchanged will suffice, or no accounting is required at all.

---

[14] Guide to Microforms in Print. München: K.G. Saur, 1978-; Guide to Microforms in Print: Supplement. Westport, CT: Microform Review, 1979-; http://www.readex.com; http://www.idc.nl

ORGANISATION AND METHODS

Ed. by Anneli Virtanen

The organisation of exchanges has changed considerably, when computers have been taken in common use. The earlier used standard forms have in many cases been compensated by computer programmes and e-mail. Nevertheless, it is good to follow some general and in common accepted practices.

While there is no doubt that the exchange service of a library should work closely with the other services of that library, and in particular with the acquisitions department, the staff employed in exchange work should be relatively independent and above all have an approach to their work, which differs from that governing acquisitions through purely commercial channels. Their aim should be to enrich the library for which they are working, while bearing in mind the legitimate interest of the foreign partners. The staff of the exchange department should have a good knowledge of languages and should take a particular interest in their rather specialized activity.

It would be a mistake to think that exchanges with libraries constitute a kind of impersonal mechanism and can be carried out purely mechanically. Exchange operations necessitate a lot of work. Exchange cannot be looked upon as a friendly correspondence between public-spirited librarians. It requires an administrative organization, even on a small scale, in order to ensure smooth operations. It is further recommended that different matters to be dealt with separately in correspondence: to mix offers, requests and inquiries for periodicals in a single letter, for example, will delay the solution of these various problems if they have to be handled by different people or even different departments.

1. Establishment of New Exchange Relations

The letter addressed to a national centre or a library, with which no exchange relations have yet been established will include several items of information: publications which the requesting library wishes to receive, publications that it offers in exchange, and the system proposed for balancing the exchanges.

If two parties reach an agreement on the three above-mentioned points, the letters exchanged between them usually constitute an 'exchange agreement' sufficient to permit the operations to continue without renewing the agreement each time. Exchanges of official publications should be considered separately if the establishment of an agreement requires action by government authorities.

The introductory letter may contain offers of publications. If the two parties have already so agreed, the publications may be sent at varying intervals, in a way that may or may not be related to the balance of their exchanges. Such offers have hitherto taken very varied forms, including lists arranged alphabetically or by subject, card indexes, catalogues and prospectuses, titles mentioned in a letter, etc. The first two forms are the best.

A distinction should be drawn, in lists of proposals, between monographs and periodicals. When a large number of works is proposed, it may be convenient to list them by subject-matter or even to draw up lists for individual subjects (literature, medicine, etc.). It is also good to mark offers with numbers or other codes, which can be used in requests. The circulation of mimeographed lists offers a great advantage to anyone proposing publications, since it enables him to reach many partners at once. The negative aspect of this method is that there might be several libraries, which are interested, in the same titles. Checking and requesting may cause lot of work in vain.

2. Replies to Offers

Proposal forms and lists can be returned to the sender after checking, or only codes are used in replies. However, replies sometimes take the form of a new list of works required, together with reference to the proposals that have been made.

Requests are submitted at present in many different ways. E-mail is often used in correspondence instead of written letters and standard forms.

3. Replies to Requests

The sending of the work requested is often a sufficient reply to the request. If it is not possible to send the work, the appropriate reply may be checked on the request form or give answers some other way. When requests from libraries are made on the basis of lists and there are not enough copies available to meet all the requests, several systems are used to cope with this situation. None of these can of course satisfy all the requesting parties. Requests may be met in the order in which they are received, in accordance with the fields in which the libraries specialize, or taking into account the balance of exchange with each of the requesting libraries.

4. Inquiries

Inquiries may have to do either with monographs or serial publications. For monographs, the inquiry may take the form of a new request form bearing the reference (including the date) of the first request. For serial publications, an international standardization would be desirable in such a way as to establish in the inquiry a distinction between cases involving a missing issue and those in which the series seems to have been stopped. The most frequent reasons for not complying with a request -delayed publication, publications stopped, change of title, etc. - would be foreseen in the replies.

5. Dispatching

Dispatching operations naturally require material organization, the extent of which will vary with the volume of the exchanges. Some libraries, which carry on limited exchanges, have their material sent by the printer in cases where the library is responsible for its publication. In some cases publishers or bookshops perform this work for a commission. In most cases, however, the libraries themselves do the sending and must employ staff for the work of packing and dispatching. Consignments should go out as quickly as possible, especially where periodicals are involved. It will be advantageous to send bulk consignments when they are not urgent, but the interval between consignments should not exceed four weeks.

6. Sending of Publications

As a rule, no publication should be sent without the addressee's agreement. What we are concerned with here, therefore, is a work that has been requested, one that is part of a series or an issue of a periodical, on which prior agreement has been reached. In some cases, by mutual agreement, certain types of publication will be sent automatically, as for example bibliographies, dictionaries and biographies, the selection being left to the sender.

In this connection is recommended the use of a dispatch note, which should accompany any individual dispatch of publications or bulk consignment, but need only be sent once a year for each periodical title (possibly with an indication of the yearly subscription for the periodical).

Each dispatch note should be dated and numbered, indicated by the number of books and the value of each of them according to the balance system that has been chosen. This information is sent to the consignee, one by letter or e-mail (announcing the dispatch) and the other with the publications themselves. One of the two copies may be used as a receipt form by being returned to the sender.

7. Recording of Exchanges

It is essential in all cases to record both the dispatches sent to a partner and those received from him. In practice, the best system is to draw up a chronological table with a parallel listing of the various consignments. The lists can be simplified if dispatch notes are used; it will then suffice to indicate the number of the note (with its date), the number of volumes and possibly their value or the number of pages they contain. It will not be necessary to repeat in detail the contents of each consignment. If it has been agreed with a partner that exchanges should balance out exactly, this system will make it possible to draw up a double-entry annual record. A simple balance sheet can thus be made up and sent to the partner for approval or correction. In any case, each partner should draw up his own balance sheet.

8. Practical Organisational Procedures

Every exchange centre will need the supplies listed briefly below for the series of operations just described.

* Files containing the correspondence exchanged with each partner, which may be classified by country, town or partner.

* If dispatch notes are used, a double file will be needed for each partner and will contain the notes for works received and works sent out. These files may be combined with the previous ones.

* An index of partners, carefully kept up to date, with the name and exact address of each of them. An acronym may also be used on various documents to identify the partner quickly.

* An index of works requested for exchange purposes.

* An index of available works, which might also contain a list of works requested by partners but temporarily unavailable.

Some libraries exchange their duplicates by the 'open house' system, which consists in allowing librarians who are interested to come and choose for themselves.

* The exchange of periodicals and serial publications has the advantage of operating almost automatically once the titles have been chosen and accepted by the partners. However, this calls for a different type of account keeping from that used for monographs. For each title there must be an index to check each issue sent and received.

* Provision should also be made for a system making it possible, whenever necessary, to draw up a balance sheet of exchanges for each partner. The importance of this system will depend upon the number of partners and the volume of exchanges with each of them. Such registration will also make it possible to draw up statistics.

* Annual statistics. It will be found advisable to draw up annual statistics in order to determine changes in the volume of operations from year to year.

* Storage. Whatever the extent of the exchanges, room must be provided to store material. A maximum storage time should be established in order to avoid con-

gestion which would jeopardize the efficiency of the system. Works may be kept from two to five years, depending upon local conditions.

9. Transport

The more smoothly the practical machinery involved in exchange works, the more adequately will the fundamental aim of the exchange of publications - to promote the flow of ideas between the scientific institutions of the various countries - be achieved. This also applies to the technical aspect of exchanges, namely, the dispatch of the exchange material.

If it takes a dispatchment a long time to reach its destination, it arrives late or irregularly, or is incomplete when it comes to hand, the exchange fails of its purpose. If the sender must comply with complicated and costly formalities in order to obtain an export licence, the exchange risks falling through. Exchanges are possible only if dispatch is easy and delivery is punctual and free from all formalities. These conditions should be self-evident; but unfortunately, the times when they were so, are long since past. There are different barriers between the various countries, barriers that can be removed only gradually.

Publications can be dispatched in three ways: by parcel post, as freight, and as printed matter, where it is possible. None of these methods enjoys absolute preference. Here, as in all exchange operations, general rules give way to individual decisions in individual cases. Sometimes, higher costs must be accepted, because speed of exchange is the main consideration. This applies to material of 'current interest', for which the direct form of exchange is to be preferred, i.e. current serial publications and journals. Exchanges of this type of material are only useful if the most recent issues reach the addressees as soon as possible after publication.

On the other hand, there are cases in which exchanges are possible only if they can be carried out with a minimum of expense, as far as technical arrangements are concerned. Here, the decision has to be left to the exchange bureaux, which wish to use the cheapest method of dispatch, regardless how long the material may take to reach its destination.

a) Postal Dispatches

In order to enable the exchange services to effect postal dispatches and in this way to promote the exchange of publications, Unesco has endeavoured to secure postal franchise for all exchange material. So far, these efforts have proved fruitless except in the case of literature for the blind for which, since September 1957, exemption from all postal taxes and registration changes has been granted at the World Postal Congress. It seems also unlikely that these efforts will be any more successful in the foreseeable future. In some countries the full postal taxes of literature for the blind have been put to use again.

In the Sixth Additional Protocol to the Constitution of the Universal Postal Union (UPU) (http://www.upu.int) in Beijing, 1999, postal items are divided in two groups: letter-post and parcel-post. Letter-post items are classified according to two systems. The first system is based on the speed of treatment of the items and the second system on the contents of the items. In the former system items are divided into priority and non-priority groups. Priority items are conveyed by the quickest route with priority and non-priority items (economy) have a lower rate but a longer delivery time. Both groups have weight limits: 2 kilograms in general, but 5 kilograms in relations between administrations admitting such items from their customers, 5 kilograms for items containing books and pamphlets, 7 kilograms for literature for the blind.

Items based on the contents are e.g. printed papers, literature for the blind and small packets. The weight limits are the same as above. It must be noticed that some countries have made decisions not allow heavier letter-post than 2 kilograms.

b) Printed Matter

In the Sixth Additional Protocol there are mentioned some reductions, which postal administrations can allow for letter-post items. For newspapers and periodicals published in the sender's country, a reduction of not more than 50 % is allowed. The same reduction can be given to books and pamphlets, music scores and maps, provided they contain no publicity matter or advertisement other than that appearing on the cover or on the fly leaves.

Printed matter category is still used in many countries. In some countries it is allowed to all but in some countries an agreement with sender organization is demanded. To this category the biggest reductions are the same as mentioned above. Printed matter covers any type of reproduction, which is wholly printed upon paper, card or other materials commonly, used in printing. Printed matter cannot be sealed with tape. Envelopes should be sealed with staples and parcels with a string. There are also limitations on maximum weight.

As printed matter post undoubtedly represents the fastest method of dispatch, it is obviously the method most indicated the material of 'current interest'. Unfortunately all the UPU member countries do not accept the discount for printing matter. The discount percentage can differ in different countries.

c) Parcel Post

In the case of direct exchanges between institutions, parcel post comes into consideration only if considerable material is dispatched to the same addressee; this occurs chiefly when the institutions concerned issue large quantities of publications, and also when the material exchanged consists of duplicates and theses.

Exchange centres nearly always have to dispatch considerable quantities of material. In order to limit their expenses, individual institutions and exchange centres must carefully compare the parcel post rates and freight rates charged for dispatches to various countries. According to the decision of UPU maximum individual weight of a parcel is 50 kilograms, but to transport even a parcel of 20 kilograms is optional. The maximum weight differs in various countries. It would be good to find out beforehand the weight limits in the receiving country.

d) Airmail

The airmail rates are still so high for many countries that this method of dispatch cannot be used for all urgent material. Nevertheless, even in this field certain reductions have been made. Under the new provisions of the UPU Congress, the same reductions shall apply to printed material conveyed by air as by surface post. To make use of all available facilities, institutions taking part in exchange should study existing rates carefully.

e) Dispatch at Freight Rate

Dispatches at freight rate are often, but not always, the cheapest means of sending and are thus preferred by exchange centres. All exchange services or institutions wishing to send their publications at freight rates - by rail, air or sea - should first consult the air, maritime or railway agencies to obtain full information on rates, formalities to be fulfilled by the sender and method of labelling and packing the parcels. Regulations on these matters vary from country to country, and the sender cannot be sure that his parcels will reach their destination safely and promptly unless all formalities for the export of publications are duly completed. In some cases it is possible to get reductions but then an agreement is needed.

10. Customs

Not merely high dispatch rates, but customs formalities, taxes, fees and other charges are among the main trade barriers to knowledge which Unesco is endeavouring to have removed. The Agreement on the Importation of Educational, Scientific and Cultural Materials was unanimously adopted by Unesco's General Conference 1950 and it has been in force since 21 May 1952.

Of this agreement, which covers all kinds of educational, scientific and cultural material, Articles I and IV are of particular importance for the exchange of publications. Article I obliges the Contracting States 'not to apply customs duties or other charges on, or in connection with, the importation...'. This is a great improvement upon the earlier formulas usually found according to which customs franchise was extended only to consignments for public scientific institutions. In many countries, the majority of scientific societies, associations and similar institutions are private organizations and as such had to pay customs duties on their imports, although these were destined for purely scientific use. In 1976 the General Conference of Unesco adopted a protocol to this agreement, which aimed to facilitate the importation of objects of an educational, scientific or cultural nature by libraries of all kinds.

Article IV does not oblige the Contracting States to adopt any specific measures, but indicates the general policy, which should be followed in official quarters with regard to exchanges:

The Contracting States undertake that they will as far as possible: (a) continue their common efforts to promote by every means the free circulation of educational, scientific or cultural materials, and abolish or reduce any restrictions to that free circulation which are not referred to in this Agreement; (b) simplify the administrative procedure governing the importation of educational, scientific or cultural materials; (c) facilitate the expeditious and safe customs clearance of educational, scientific or cultural materials.

Even here, the UPU Congress of 1957 made some progress towards the free flow of information by the virtual abolition of customs clearance charges on books and other printed matter not liable to import duties and by urging member countries not to impose customs charges on books, newspapers, magazines and book catalogues.

a) Customs Formalities

Experience shows that senders of exchange material often fail to comply with customs regulations, with the result that the addressee has to pay higher customs taxes and duties than necessary.

Below is given a summary of the formalities, which the sender must complete in order to ensure that the addressee will be able to clear the publications with the customs authorities with the minimum delay.

b) General Formalities

For each parcel sent, the sender must fill in one or more customs forms, stating precisely the nature, quantity, the gross weight, country of origin, and articles with their price contained therein. In the most countries the dispatch note abroad includes the custom declaration. Certain countries require in addition an invoice or a consular declaration. Sometimes this information has to be sent beforehand to the recipient. In the event of failure to supply one of these details, the customs authorities of the receiving country are entitled to exact the maximum rate and may even impose a fine on the addressee or the sender.

c) Number of Customs Declarations Demanded

The number of customs declarations demanded varies very much: while certain countries require one only, others require two, three or even four. The sender would do well to inquire either of the local transport agency or of the foreign addressee how many customs declarations are required by the country, to which the parcel is being sent. If an item other than a letter is sent as letter-post or small packet, a green 'customs' label on the envelope is demanded.

d) Import and Export Licences

There are a few countries which require import licences for publications; but as a general rule licences are requires only for export, in particular in countries subject to exchange control. These licences are, however, not difficult to obtain in respect of material exchanged between scientific, cultural or university institutions.

We can, however, hope that the path of institutions engaging in exchanges will gradually be smoothed, and that the agreement will enable 'the free flow of ideas by word and image' to become a reality in the sphere of the international exchange of publications.

INTERNATIONAL FRAMEWORK OF EXCHANGES

Ed. by Pentti Vattulainen

1. Conventions and Agreements for the Exchange of Publications

The fourth edition of the handbook gives a survey of the different forms in which exchange agreements can be concluded.

There have been

(a) multilateral conventions between governments on the basis of which the member countries enter into exchange relations together — generally by way of exchange centres;

(b) bilateral arrangements for the exchange of publications whether informal ones by exchange of notes or letters, or by formal texts;

(c) general cultural agreements (whether bilateral or multilateral) with a clause referring directly to exchange or, in general, to the dissemination of scientific literature and by which exchange relations between institutions of the contracting parties may be stimulated or even initiated.

The only conventions that may still have validity are the two Unesco conventions of 1958. The two conventions: the Convention Concerning the International Exchange of Publications (Convention 1) and the Convention Concerning the Exchange of Official Publications and Government Documents Between States (Convention 2) were unanimously approved by the tenth session of Unesco's General Conference held in Paris in 1958 and consequently opened for ratification by Member States. Both conventions came into force in 1961, Convention 1 on 23 November and Convention 2 on 30 May. The text of both conventions is given in this chapter.

2. Texts of Intergovernmental Agreements for the International Exchange of Publications

2.1. Convention Concerning the International Exchange of Publications, Adopted by the General Conference at its Tenth Session, Paris, 3 December 1958

Text of the Convention

The General Conference of the United Nations Educational, Scientific and Cultural Organization, meeting in Paris from 4 November to 5 December 1958, at its tenth session, Convinced that development of the international exchange of publications is essential to the exchange of ideas and knowledge among the peoples of the world. Considering the importance accorded to the international exchange of publications by the Constitution of the United Nations Educational, Scientific and Cultural Organization,

(Original text in French.)

Recognizing the need for a new international convention concerning the exchange of publications,

Having before it proposals concerning the international exchange of publications constituting item 15.4.1 on the agenda of the session,

Having decided, at its ninth session, that these proposals should be made the subject of international regulation by way of an international convention,

Adopts, this third day of December 1958, the present Convention.

article 1. Exchange of Publications

The Contracting States undertake to encourage and facilitate the exchange of publications between both governmental bodies and non-governmental institutions of an educational, scientific and technical, or cultural nature, which are non-profit-making in character, in accordance with the provisions of the present Convention.

article 2. Scope of the Exchange of Publications

1. For the purpose of the present Convention, the following publications may be considered appropriate articles to be exchanged, for use but not for resale, between the bodies and institutions referred to in Article 1 of the present Convention: (a) publications of an educational, legal, scientific and technical, cultural and informational nature such as books, newspapers and periodicals, maps and plans, prints, photographs, microcopies, musical works, Braille publications and other graphic material; (b) publications covered by the Convention Concerning the Exchange of Official Publications and Government Documents between States, adopted by the General Conference of the United Nations Educational, Scientific and Cultural Organization on the third day of December 1958.

2. The present Convention in no way affects exchanges carried out under the Convention concerning the Exchange of Official Publications and Government Documents between States, adopted by the General Conference of the United Nations Educational, Scientific and Cultural Organization on the third day of December 1958.

3. The present Convention does not apply to confidential documents, circulars and other items which have not been made public.

article 3. Exchange Services

1. The contracting States may entrust the national exchange service or where no such national exchange service exists, the central exchange authority or authorities with the following functions in connection with the development and coordination of the exchange of publications among bodies and institutions referred to in Article 1 of the present Convention:

a. facilitating the international exchange of publications, in particular by transmitting, when appropriate, the material to be exchanged;

b. supplying advice and information on exchange possibilities for bodies and institutions at home and abroad;

c. encouraging, when appropriate, the exchange of duplicate material.

2. However, when it is deemed undesirable to centralize in the national exchange service or in central authorities the development and co-ordination of exchanges among bodies and institutions referred to in Article 1 of the present Convention, any or all of the functions enumerated in paragraph 1 of the present article may be entrusted to other authority or authorities.

article 4. Method of Transmission

The transmission may be made either directly between the bodies and institutions concerned, or through the national exchange service or exchange authorities.

article 5. Transport Charges

When transmissions are made directly between exchange partners, the Contracting States shall not be required to bear the cost thereof. If the transmission is made through the exchange authority or authorities, the Contracting States shall bear the cost of the transmission as far as destination, but, for transport by sea, the cost of packing and carriage shall be paid only as far as the customs office of the port of arrival.

article 6. Rates and Conditions of Transport

The Contracting States shall take all the necessary measures to ensure that the exchange authorities benefit from the most favourable existing rates and transport conditions, whatever the means of transport chosen: post, road, rail, inland or sea transport, airmail or air cargo.

article 7. Customs and Other Facilities

Each contracting State shall grant its exchange authorities exemption from customs duties for both imported and exported material under provisions of the present Convention or under any agreement in implementation thereof and shall accord them the most favourable treatment as regards customs and other facilities.

article 8. International Co-ordination of Exchange

To assist the United Nations Educational, Scientific and Cultural Organization in the performance of the functions assigned to it by its Constitution concerning the international co-ordination of exchange, the Contracting States shall send to the Organization annual reports on the working of the present Convention and copies of bilateral agreements entered into in accordance with Article 12.

article 9. Information and Studies

The United Nations Educational, Scientific and Cultural Organization shall publish information received from the Contracting States in application of Article 8 and shall prepare and publish studies on the working of the present Convention.

article 10. Assistance of Unesco

1. The Contracting States may call upon the United Nations Educational, Scientific and Cultural Organization for technical assistance in connection with any problem arising out of the application of the present Convention. The Organization shall accord such assistance within the limits fixed by its programme and its resources, in particular for the creation and organization of national exchange services.

2. The Organization is authorized to make, on its own initiative, proposals on this matter to the Contracting States.

article 11. Relation to Previous Agreements

The present Convention shall not affect obligations previously entered into by the Contracting States by virtue of international agreements.

article 12. Bilateral Agreements

Whenever necessary or desirable, the Contracting States shall enter into bilateral agreements for the purpose of supplementing the present Convention and regulating matters of common concern arising out of its application.

article 13. Languages

The present Convention is drawn up in English, French, Russian and Spanish, the four texts being equally authoritative.

article 14. Ratification and Acceptance

1. The present Convention shall be subject to ratification or acceptance by States members of the United Nations Educational, Scientific and Cultural Organization, in accordance with their respective constitutional procedures.

2. The instruments of ratification or acceptance shall be deposited with the Director-General of the United Nations Educational, Scientific and Cultural Organization.

article 15. Accession

1. The present Convention shall be open for accession by all States not members of the Organization invited to do so by the Executive Board of the United Nations Educational, Scientific and Cultural Organization.

2. Accession shall be effected by the deposit on an instrument of accession with the Director-General of the United Nations Educational, Scientific and Cultural Organization.

article 16. Entry into Force

The present Convention shall enter into force twelve months after the date of the deposit of the third instrument of ratification, acceptance or accession, but only with respect to those States which have deposited their respective instruments on or before that date. It shall enter into force for each other State which deposits its instrument of ratification, acceptance or accession, twelve months after the deposit of its instrument of ratification, acceptance or accession.

article 17. Territorial Extension of the Convention

Any Contracting State may, at the time of ratification, acceptance or accession, or at any time thereafter, declare by notification addressed to the Director-General of the United Nations Educational, Scientific and Cultural Organization that the present Convention shall extend to all or any of the territories for whose international relations it is responsible. The said notification shall take effect twelve months after the date of its receipt.

article 18. Denunciations

1. Each Contracting State may denounce the present Convention on its own behalf or on behalf of any territory for whose international relations it is responsible.

2. The denunciation shall be notified by an instrument in writing, deposited with the Director-General of the United Nations Educational, Scientific and Cultural Organization.

3. The denunciation shall take effect twelve months after the receipt of the instrument of denunciation.

article 19. Notifications

The Director-General of the United Nations Educational, Scientific and Cultural Organization shall inform the States members of the Organization, States not members of the Organization referred to in Article 15, as well as the United Nations, of the deposit of all the instruments of ratification, acceptance and accession provided for in Articles 14 and 15 and of the notifications and denunciations provided for respectively in Articles 17 and 18.

article 20. Revision of the Convention

1. The present Convention may be revised by the General Conference of the United Nations Educational, Scientific and Cultural Organization, any such revision, however, binding only those States which shall become parties to the revising convention.

2. Should the General Conference adopt a new convention revising the present Convention in whole or in part, and unless the new convention otherwise provides, the present Convention shall cease to be open to ratification, acceptance or accession as from the date when the new revising convention enters into force.

article 21. Registration

In accordance with Article 102 of the Charter of the United Nations, the present Convention shall be registered with the Secretariat of the United Nations at the request of the Director-General of the United Nations Educational, Scientific and Cultural Organization.

2.2. Convention Concerning the Exchange of Official Publications and Government Documents Between States, Adopted by the General Conference at its Tenth Session, Paris, 3 December 1958

Text of the Convention

The General Conference of the United Nations Educational, Scientific and Cultural Organization, meeting in Paris from 4 November to 5 December 1958, at its tenth session,

Convinced that development of the international exchange of publications is essential to the free exchange of ideas and knowledge among the peoples of the world,

Considering the importance accorded to the international exchange of publications by the Constitution of the United Nations Educational, Scientific and Cultural Organization.

Being aware of the provisions for the exchange of official publications set forth in the Convention for the International Exchange of Official Documents, Scientific and Literary Publications and in the Convention for the Immediate Exchange of Official Journals, Public Parliamentary Annals and Documents, concluded in Brussels on 15 March 1886 and in various regional agreements for the exchange of publications,

Recognizing the need for a new international convention concerning the exchange of official publications and government documents between States,

Having before it proposals concerning the exchange of official publications and government documents between States constituting item 15.4.1 on the agenda of the session,

Having decided, at its ninth session, that these proposals should be made the subject of international regulation by way of an international convention,

Adopts, this third day of December 1958, the present Convention.

article 1. Exchange of Official Publications and Government documents

The contracting States express their willingness to exchange their official publications and government documents, on a reciprocal basis, in accordance with the provisions of the present Convention.

article 2. Definition of Official Publications and Government Documents

1. For the purpose of the present Convention, the following are considered official publications and government documents when they are executed by the order and at the expense of any national governmental authority: parliamentary documents, reports and journals and other legislative papers; administrative publications and reports from central, federal and regional governmental bodies; national bibliographies, State handbooks, bodies of law, decisions of the Courts of Justice; and other publications as may be agreed.

2. However, in the application of the present Convention, the Contracting States shall be free to determine the official publications and government documents which shall constitute exchange material.

3. The present Convention, does not apply to confidential documents, circulars and other items which have not been made public.

article 3. Bilateral Agreements

The Contracting States, whenever they deem it appropriate, shall enter into bilateral agreements for the purpose of implementing the present Convention and regulating matters of common concern arising out of its application.

article 4. National Exchange Authorities

1. In each Contracting State, the national exchange service or, where no such service exists, the central authority or authorities designated for the purpose, shall carry out the functions of exchange.

2. The exchange authorities shall be responsible within each Contracting State for the implementation of the present Convention and of bilateral agreements as referred to in Article 3, whenever appropriate. Each Contracting State shall give its national exchange service or the central exchange authorities the powers required to obtain the material to be exchanged and sufficient financial means to carry out the functions of exchange.

article 5. List and Number of Publications for Exchange

The list and number of official publications and government documents for exchange shall be agreed between the exchange authorities of the Contracting States. This list and the number of official publications and government documents for exchange may be modified by arrangements between such authorities.

article 6. Method of Transmission

Transmissions may be made directly to exchange authorities or to recipients named by them. The method of listing consignments may be agreed between exchange authorities.

article 7. Transport Charges

Unless otherwise agreed, the exchange authority which undertakes the transmission shall bear the cost thereof as far as destination, but for transport by sea, the cost of packing and carriage shall be paid only as far as the customs office of the port of arrival.

article 8. Rates and Conditions of Transport

The Contracting States shall take all the necessary measures to ensure that the exchange authorities benefit from the most favourable existing rates and transport conditions, whatever the means of transport chosen: post, road, rail, inland or sea transport, airmail or air cargo.

article 9. Customs and Other Facilities

Each Contracting State shall grant its exchange authorities exemption from customs duties for both imported and exported material under the provisions of the present Convention or under any agreement in implementation thereof and shall accord them the most favourable treatment as regards customs and other facilities.

article 10. International Co-ordination of Exchange

To assist the United Nations Educational, Scientific and Cultural Organization in the performance of the functions concerning the international co-ordination of exchange assigned to it by its Constitution, the Contracting States shall send to the Organization annual reports on the working of the present Convention and copies of bilateral agreements entered into in accordance with Article 3.

article 11. Information and Studies

The United Nations Educational, Scientific and Cultural Organization shall publish information received from the Contracting States in conformity with Article 10 and shall prepare and publish studies on the working of the present Convention.

article 12. Assistance of Unesco

1. The Contracting States may call upon the United Nations Educational, Scientific and Cultural Organization for technical assistance in connection with any problem arising out of the application of the present Convention. The Organization shall accord such assistance within the limits fixed by its programme and its resources, in particular, for the creation and organization of national exchange services.

2. The Organization is authorized to make, on its own, proposals on these matters to the Contracting States.

article 13. Relation to Previous Agreements

The present Convention shall not affect obligations previously entered into by the Contracting States by virtue of international agreements. It shall not be construed as requiring a duplication of exchanges conducted under existing agreements.

article 14. Languages

The present Convention is drawn up in English, French, Russian and Spanish, the four texts being equally authoritative.

article 15. Ratification and Acceptance

1. The present Convention shall be subject to ratification or acceptance by States members of the United Nations Educational, Scientific and Cultural Organization, in accordance with their respective constitutional procedures.

2. The instruments of ratification or acceptance shall be deposited with the Director-General of the United Nations Educational, Scientific and Cultural Organization.

article 16. Accession

1. The present Convention shall be open for accession by all States not members of the Organization invited to do so by the Executive Board of the United Nations Educational, Scientific and Cultural Organization.

2. Accession shall be effected by the deposit of an instrument of accession with the Director-General of the United Nations Educational, Scientific and Cultural Organization.

article 17. Entry into Force

The present Convention shall enter into force twelve months after the date of the deposit of the third instrument of ratification, acceptance or accession, but only with respect to those States which have deposited their respective instruments on or before that date. It shall enter into force for each other State which deposits its instrument of ratification, acceptance or accession, twelve months after the deposit of its instruments of ratification, acceptance or accession.

article 18. Territorial Extension of the Convention

Any Contracting State may, at the time of ratification, acceptance or accession, or at any time thereafter, declare by notification addressed to the Director-General of the United Nations Educational, Scientific and Cultural Organization that the present Convention shall extend to all or any of the territories for whose international relations it is responsible. The said notification shall take effect twelve months after the date of its receipt.

article 19. Denunciation

1. Each Contracting State may denounce the present Convention on its own behalf or on behalf of any territory for whose international relations it is responsible.

2. The denunciation shall be notified by an instrument in writing, deposited With the Director-General of the United Nations Educational, Scientific and Cultural Organization.

3. The denunciation shall take effect twelve months after the receipt of the instrument of denunciation.

article 20. Notifications

The Director-General of the United Nations Educational, Scientific and Cultural Organization shall inform the States members of the Organization, States not members of the Organization referred to in Article 16, as well as the United Nations, of the deposit of all the instruments of ratification, acceptance and accession provided for in Articles 15 and 16 and of the notifications and denunciations provided for respectively in Articles 18 and 19.

article 21. Revision of the Convention

1. The present Convention may be revised by the General Conference of the United Nations Educational, Scientific and Cultural Organization, any such revision, however, binding only those States which shall become parties to the revis-

ing convention.

2. Should the General Conference adopt a new convention revising the present Convention in whole on in part, and unless the new convention otherwise provides, the present Convention shall cease to be open to ratification, acceptance or accession as from the date when the new revising convention enters into force.

article 22. Registration

In accordance with Article 102 of the Charter of the United Nations, the present Convention shall be registered with the Secretariat of the United Nations at the request of the Director-General of the United Nations Educational, Scientific and Cultural Organization.

3. The Role of Unesco

As the importance of exchanges has diminished, the need for Unesco's involvement has disappeared ad Unesco nowadays has no activities going on with the exchanges.

4. The Role of IFLA

The International Federation of Library Associations and Institutions (IFLA) was founded in 1927. Its aim is to promote co-operation in the field of librarianship and bibliography, and particularly to carry out investigations and make proposals concerning international relations between libraries, library associations, bibliographers, and other organized groups.

IFLA is an international non-governmental organization with all the advantages and disadvantages inherent in such a constitution: on the one hand, independent without obligation towards any government but on the other hand, sometimes having difficulty in inducing the State governments of its member associations to carry out the resolutions adopted in the plenary sessions of the annual conferences attended by a great number of participants.

The proceedings of the annual conferences give a clear insight in the work of IFLA which has developed into a world-wide organization with ever-increasing activities.

IFLA carries out its activities through sections and exchanges fall into the scope of Acquisitions and Collection Development Section of IFLA.

PART TWO: HISTORY AND EXAMPLES OF OUR DAYS

HISTORY OF THE INTERNATIONAL EXCHANGE OF PUBLICATIONS

Johanna Lilja
The National Board of Antiquities Library
Finland

Many proposals have been made that a history of the international exchange of publications should be written. No such history has been published, only a few articles.[1] The exchange of publications is connected with many other things: development of science and scientific publishing, library policy, diplomatic relations and international co-operation. Its history should be examined taking notice of all these activities. This brief chapter does not aim to cover the whole history of exchange, but to describe the main points in its development. Neither is this text based on archival sources. The most important sources have been the previous *Handbooks on the International exchange of publications* published by Unesco in 1950, 1956, 1964 and 1978 and papers from the congresses concerning exchange. This chapter ends in the 1970's. Since the 1980's new issues have arisen. Many of them are discussed in other chapters of this handbook.

Origins of the Exchange of Publications

Already the medieval libraries sometimes exchanged manuscripts, but the systematic history of exchange of publications begins at the dawn of scientific publishing. In the 17th century publishing scientific papers and exchanging them was in the most cases the same thing. The correspondence between scholars or

---

[1] In the Resolutions of the Conference of the International Exchange of Publications in Europe (Budapest 1960) there was a recommendation to Unesco to undertake studies on the theory and historical development of exchange of publications. *Conférence sur les échanges internationaux de publications en Europe. Budapest, 13-19 septembre 1960*, Budapest, 1962, p. 241 ; Also Miss Dargent, the compiler of the bibliography of the international exchange of publications, proposed that a history of international exchange of publications should be written. István Gombocz, The forty Years of the Committee on the Exchange of Publications, edited by Maria J. Schiltman, *IFLA Journal*, (Pullach-München),Vol. I, 1974, p. 18 ; The most important article concerning the history of the exchange of publications has been: Sarah S. Gibson, Scientific Societies and Exchange : a Facet of the History of Scientific Communication, *Journal of Library History*, Austin (MN), Vol. 17, No 2, Spring 1982, p. 144. For writing this brief history I have received valuable help from D. Soc. Sci. Ilkka Mäkinen, my colleague, librarian Päivi Myllykoski and the personnel of the Exchange Centre for Scientific Literature.

between scientific societies is a common root both to the exchange of publications and to scientific periodicals.

Scientific societies emerged in the ideological climate of the Renaissance. Among the first societies was *Accademia dei Lincei,* founded in Rome in 1603 by Prince Federico Cesi. Its purpose was "the acquisition of wisdom and knowledge of things" and "the announcement of these to men". During the 17th century many scientific societies and academies were founded in Europe to promote scientific research – to create knowledge of nature by experiments and inductive reasoning. New national states gave economical support to these societies and academies. The establishment of reliable and regular postal services in Europe made it possible to organise a network of scientific correspondence. (Gibson 1982, 145 ; Manten 1980, 4 ; Broman 2000, 226-227.)

The scholars used correspondence to report on their research or to request aid in gathering or securing information, for example in collecting astronomical observations from different parts of Europe. This kind of correspondence was not usually of a private character. On the contrary, it was rather meant to convey reports to the community of scholars. The reciprocity between correspondents was considered very important; a letter for a letter, news for news, a favour for a favour. Sometimes the correspondence was printed. For instance *Accademia dei Lincei* published Galileo's letters on sunspots in the monograph *Macchie solari* in 1612. Printing a monograph was, however, expensive, especially if pictures were used, and it certainly was not the most convenient way to convey observations to colleagues. It was more practical to make the correspondence more systematic. The letters were transmitted through an intermediary who copied and distributed them to other scholars or scientific societies. The most famous of intermediaries was Henry Oldenburg, the Secretary of the Royal Society, who organized the correspondence between the society and the scholarly community in Continental Europe. (Gibson 1982, 146-148 ; Manten 1980, 4 ; Broman 2000, 228 ; Goldgar, 1995, 21-26.)

Copying letters became quite a heavy task, because new scientific societies were founded throughout the 17th century. Besides, the letters did not guarantee claims for priority – someone else could publish the information of a letter. To simplify his correspondence and also to make a little money Henry Oldenburg inaugurated in 1665 the journal *Philosophical Transactions : Giving Some Accompt of the Present Undertakings, Studies and Labours of the Ingenious in Many Parts of the World.* In the same year also appeared the first number of *Journal des Sçavans* founded by the French jurist and aristocrat Denis de Sallo. In the following decades many other journals emerged: *Giornale de Litterati d'Italia* (1668), *Miscellanea Curiosa* (1670), *Acta Medica et Philosophia Hafniensia* (1673), *Acta Eruditorum* (1682) and *Nouvelles de la République des Lettres* (1684). (Broman 2000, 228-229 ; Manten 1980, 5-9). Scientific journals did not end the correspondence and the scholarly system of mutual aid. Rather, they grew attached to this informal network. In the Seventeenth century the book trade could not satisfy the needs of the readers. Hence the international community of men of letters helped each other by sending books and journals to colleagues abroad. (Goldgar 1995, 16-17, 26.)

As the scientific publishing increased, the societies aspired to found their own libraries. *Accademia dei Lincei* appointed a librarian, whose primary duty was presumably the distribution of gift copies of publications and manuscripts. This gift-giving was thought to be reciprocal in nature and may thus be considered as a forerunner of the exchange of publications. In spite of these efforts the society was not successful in assembling a library. (Gibson 1982, 146.) The Royal Society, like many other English societies, had a library of its own from the beginning. Its collections grew mostly by the donations of the members (Wyatt 1997, 191-194). In Germany scientific societies had their own libraries from the end of the 18th century. Usually these were open only to the members of the society (Berninger 1997, 7-8).

In the Eighteenth century, access to scientific information was still dependent on one's membership in the scholarly community. Exchange of publications between libraries had begun, but it was not very systematic and libraries were not meant to be open to the public. Neither was the commercial publishing and marketing of scientific periodicals well organized.

Organizing Exchange in the Universities

By the time the societies began to publish periodicals and founded scientific libraries, most university libraries in Europe were very modest. The actual task of a university library in the Seventeenth century was not supporting research by acquiring current literature. These Baroque libraries were rather depositories of old and valuable books and also other kind of curiosities – monstrosities, fossils, holy relics, coins, medals and so on. Annual budgets were collected from student fines or fees and usually they were miserable. Furthermore, many libraries had suffered from the reformation and religious wars. Endowments and testaments were the most important way of acquisitions. This meant that private libraries found their way to the university libraries' shelves *en bloc*, whether these books were needed or not. Selling or exchanging duplets was one way for libraries to build their collections. (Harris 1984, 130-137 ; Clark 2000, 190-193 ; Södergård 1992, 3 ; Vallinkoski 1948, 178-181, 185-204.)

The ideas of the Enlightenment changed the library policy. Universities began to adopt new methods of scientific research, instead of citing old authorities like the Bible or Aristotle. University of Göttingen, founded in 1737, stood in the forefront representing a pragmatic, rationalising view of knowledge and in building a modern research library. Its library had sufficient funds, due to the generosity of minister Gerlach Adolph von Münchhausen. Thus the University library of Göttingen could build its collections systematically by choosing useful books and research materials. It also guaranteed the access to collections by reorganizing the catalogues. However, most of the Eighteenth century libraries did not have generous patrons like Münchhausen, and they still had to make acquisitions with very modest funds. (Kleinert 1997, 103-106 ; Clark 2000, 196-200.)

Exchanging publications was one solution to the acquisition problems of the university libraries. Swedish universities were forerunners in organizing exchange. In 1745 the Vice-Chancellor of the University of Uppsala, Jacob Benzelius suggested that in order to improve contacts between Swedish universities, thirty copies of each academic publication should be reserved for exchange. In the beginning the exchange was a business of universities, not their libraries: publications received from other universities were to be distributed to the professors. However, it was likely that even libraries got copies to their collections. This exchange circle, called *commercium literarium,* started between the universities of Uppsala, Lund, Turku and Greifswald. (All these universities were situated in Swedish territory at the time.) C*ommercium literarium* suffered from the same problems as the exchange in our days: defects in consignments, difficulties in transportation, misunderstandings etc. In spite of these difficulties, university exchange continued until the 19th and 20th centuries and extended to universities beyond the Swedish territory. Gradually, the libraries became the depositories of the exchange material and also the distribution of the publications was transferred to the library staff. (Bring 1929, 130-131 ; Vallinkoski 1975, 146-148.)

The Swedish example was soon followed in Germany. In 1817 German universities organized an association called *Akademischer Tauschverein* to promote the exchange of university publications. In the beginning the *Tauschverein* was meant to include only German universities, but soon an article in the scientific journal *Isis* spread the word of this possibility. In the next decades universities and other scientific institutions from Russia, Poland, Scandinavian countries, Netherlands, Belgium, Switzerland, Austria, Italy, England, United States and Australia joined the *Tauschverein.* By the year 1882 the number of copies, which the publisher (sometimes it was university, sometimes the poor doctoral candidate) had to deliver to the exchange rose to 50. Thus it is not surprising that eighteen French universities joining the *Tauschverein* caused the falling apart of this association. The number of exchange copies required was too high for the little universities. Nevertheless, many of the exchange relations of universities continued to the twentieth century. (Zur Geschichte des akademischen Tauschvereins 1885, 471-473 ; Jörgensen1930, 114-115 ; Bring 1929 131-132 ; *Handbook* 1978, 16-17.)

Both *Commercium Literarium* and *Akademische Tauschverein* were in the beginning confused by questions of ownership. Initiative to the exchange came usually from university professors and also the distribution was in the first decades done by the universities. Hence the material obtained through the exchange was also sometimes considered to be the private property of the professors. *(Handbook* 1978, 16-17 ; Bring 1929, 130 ; Vallinkoski 1975, 146-148.) Informal networks were still very important for young scholars if they wanted to have access to scientific information.

Exchange of Publications Personified — Alexandre Vattemare

Exchanging publications helped university libraries to acquire academic material but this was only a partial solution to the information needs of the scholars. Besides, French Revolution and the Napoleonic wars had damaged many libraries in Europe at the end of the Eighteenth and the beginning of the Nineteenth century. In the United States the situation was not any better. Fire had destroyed the modest collection of the Library of Congress in 1814. Apart from Harvard, most college libraries were small and their collections were based mostly to donations. Scientific societies had begun the exchange of publications and this innovation was also introduced to the Library of Congress. In 1834 the Congress passed a resolution providing that twenty-five copies of each work printed at government expense would be placed at the disposal of the Joint Committee of the Library to be given in return for the donations to the library. This resolution made possible the international exchange of official publications, but it did not make the exchange practice regular. A step forward was the authorization of the librarian to exchange public documents with the French government. (Harris 1984, 160-169, 183-184 ; Armbruster 1997, 133; Gibson 1982, 153.) However, there was still a lot to be done in developing exchange and making it world-wide.

The efforts to develop the exchange of publications personified in the Nineteenth century in Alexandre Vattemare, a famous French actor and ventriloquist. Touring all over Europe Vattemare established a wide range of connections with princes and courtiers as well with the bourgeois and rural public. To prepare his performances he visited the local libraries and museums and this way he observed that many libraries held valuable copies of local literature but that their foreign collections were very modest. In the beginning of his career Vattemare helped private collectors to exchange items abroad. In 1832 the librarian of the National Library of Munich asked him to arrange an exchange of the duplicates of the Munich library for fine arts materials from Paris. This meant a beginning of a new phase in Vattemare's life. (Armbruster 1997, 132-133.)

Soon Vattemare's services became well-known and he began to receive exchange lists from directors of museums all over Europe. Items offered for exchange were books and also works of art and natural phenomena. Encouraged by the support he had from many scientists, artists and government officials, even the Tsar Nicholas I of Russia, Vattemare turned to the French government. His purpose was to found an office of exchange in Paris. However, the French government was not interested in exchanging duplicates. It continued to limit its exchange to official documents and preferred to rely on diplomatic channels rather than independent agents. (Armbruster 1997, 133-134).

Vattemare did not give up. He had absorbed the Enlightenment ideal of increasing knowledge in society and in the beginning of the Nineteenth century the *République des Lettres* was not considered too large for one man. In 1839 he travelled to New York to enlist American institutions in his exchange system. He organized public meetings for propagating the idea of exchange and after few months he turned to the United States Congress proposing that the United States Government should enter his international system of exchange. He listed tens of thousands of duplicates in European libraries, including incunabulas, emphasizing that the Americans should not be worried about their short history of publishing; they could use natural specimens, fossils and their technical expertise, for example patents, in exchange – a bullfrog or a rattlesnake for the best moral and philosophical works. Vattemare's efforts succeeded. In July 1840 a bill was signed into law authorizing the Librarian of Congress to exchange duplicates held by the library, and fifty additional copies of Congressional documents were to be printed for the purpose of foreign exchange. Also several states, city corporations, educational institutions and learned societies co-operated and appointed Vattemare an agent for their exchange. He established exchanges even in Canada and Cuba. (Gibson 1982, 153-154 ; Armbruster 1997, 137-142.)

When Vattemare returned to Paris in 1841 he was in public regarded as a cultural ambassador between Europe and the United Sates and Canada. In spite of this success he did not receive support from the French government. In 1847 he travelled again to America. This time his purpose was to make an agency he had established in Paris, *Agence centrale du système d'échange international*, the official exchange agent for Congress. Vattemare brought impressive gifts from France, but when he arrived in the United States, he got an enormous bill for the duty fees of these gifts. Indefatigable Vattemare started lobbying, not only for his agency, but also for duty-free import of exchange material. In 1848 the Joint Committee on the Library appointed Vattemare its international exchange agent. Congress also granted Vattemare duty-free imports of materials arriving from Europe. Thus this talented man succeeded to overcome the custom problem, which was later to be one of the major barriers in the exchange of publications.

However, Vattemare's success did not last very long. While he was visiting the United States, a revolution occurred in France in 1848, followed by the regime of the Second Empire from 1851 on. The new French government was even less willing to co-operate with Vattemare than its predecessor. The staff of the French government approached the Librarian of Congress insisting that the exchange of official publications should be organized through diplomatic channels, not using the private services of Vattemare. In the new regime it was difficult for Vattemare to provide enough exchange material and in 1852 Congress repealed the authorizing act for his agency. Gradually Vattemare lost also the other exchange agreements with American institutions. Finally the American Civil War made reciprocal shipments almost impossible. Vattemare died in 1864 and his agency did not survive him. (Gibson 1982, 154-155 ; Armbruster1997, 143-147.)

With Alexandre Vattemare's death one phase in the exchange of publications came to its end. It was a period of informal networks and personal connections. Information had passed from one individual to another either formally published or, informally, in a letter entering either the university library or the private library of the scholar. In the second half of the Nineteenth century the volume of publishing was growing remarkably and scientific publishing began to be taken up by commercial enterprises. The network of libraries was also becoming more complicated. Apart from universities, academies and scientific societies, also museums, research institutes, laboratories and technical schools had founded specialized libraries with their own acquisition policies. Library education began, professional librarians took care of libraries and the control of the exchange became their responsibility. *République des Lettres* had enlarged so much that it was impossible for one man or one *Agence centrale* to be responsible for the exchange of publications.

The Brussels Conventions

Alexandre Vattemare was not alone with his idea of an agency attending the exchange of publications. An Englishman, James Smithson, left a large capital for an institution, which would publish its own serials and enter into correspondence with the world's learned societies. The Smithsonian Institution was founded in Washington in 1846 and it began to publish and exchange its serials three years later. This industrious institute set up agencies in different countries to provide the distribution of its publications overseas. Soon it also overtook forwarding publications from other United States institutions. It had a blessing of the Congress: in 1854 The Smithsonian Institution was given the right to ship publications duty-free and in 1867 the Congress decided to give fifty copies of its publications to be used by the Smithsonian Institution in exchange with other countries. (Gibson 1982, 155 ; *Handbook* 1964, 32 ; *Handbook* 1978, 17-18 ; Exchange of Publications 1972, 284.)

The need for official publications of other countries was recognised in the United States as well as in Europe in the last half of the 19th century. The old European class society had lost it stability, new ideologies, political upheavals, urbanisation and social problems created an order for social sciences and a need to modernise administration. Information was needed about administration, jurisdiction and statistics, not only from national but also from an international perspective.

The Smithsonian Institution had become the model for an efficiently functioning exchange agency. It seemed apparent that similar agencies were needed also in other countries. The International Congress of Geographical Sciences held in Paris in 1875 considered these problems. As a result of the proposals of this Congress, exchange services were established by 1880 in France, Portugal, Switzerland, Russia and Belgium. In Great Britain Her Majesty's Stationary Office was given responsibility for the distribution of British official publications and the British Museum responsibility for collecting and cataloguing official publications received through exchange. (Exchange of Publications 1972, 284 ; Harris 1998, 348-353.)

The first international conventions regulating the exchange of publications were created at Brussels in 1886. *Convention A for the International Exchange of Official Documents, Scientific and Literary Publications* was signed by Belgium, Brazil, Italy, Portugal, Serbia, Spain, Switzerland and the United States of America. *Convention B for the Immediate Exchange of Official Journals, Public Parliamentary Annals and Documents* was signed by Belgium, Brazil, Italy, Portugal, Serbia, Spain and the United States of America.

Convention A declared that each contracting state should establish an exchange bureau to handle the exchange of official publications – this meant official documents and works executed by the order and at the expense of the governments. Each bureau should print up to date lists of the publications available in exchange and distribute these lists to other bureaux. The exchange arrangements and shipments should be made between bureaux. Each state should assume the expenses of packing and transportation. Exchange bureaux could serve also in a non-official capacity as intermediaries between learned bodies and literary and scientific societies, but in such case their duty would be confined to the free transmission of the works exchanged. Bureaux did not have permission to take initiative in bringing about the establishment of such relations, so as to not disturb the freedom and independence of science. *Convention B* declared that the respective governments should undertake to transmit to the legislative chambers of each contracting state a copy of the official journal and of parliamentary annals and documents.[2]

One aim of the Brussels Conventions was to decentralize the responsibility of exchange to national centres and thus avoid the weakness of Vattemare's one man's scheme. Communication between countries appeared now to be duly established. However, the Brussels Conventions did not create a world-wide arrangement, because some important countries like France, Germany, Russia and Great Britain never adhered to them. (Gibson 1982, 155-156 ; Handbook 1978, 18.) In spite of this the Brussels Conventions were a crucial step in developing exchange in the end of the 19th and the beginning of the 20th century. Now there was a model for national exchange services and an international manifestation of will to co-operate in exchange. The Brussels Conventions became a shelter under which the exchange could develop even in the countries, which were not signatories. (Cox 1962, 82.)

---

[2] The text of the Brussels Conventions is published in the *Handbook on the International Exchange of Publications,* 3rd ed., Ed. Gisela von Busse, Paris, 1964, p. 61-62.

The Exchange of Publications from the Brussels Conventions to the World War II

All signatories of the Brussels Conventions, except Spain, had founded an exchange centre by the 1890's. Argentina and Paraguay adhered to the Convention in 1889 and Paraguay founded immediately a centre in connection with the Ministry of ForeignAffairs. Switzerland's national exchange centre became gradually a model to the other centres. It adhered strictly to the rules laid down by the Convention, never taking any initiative or attempting to exercise any influence on the organising of exchanges. (Handbook *1956, 238, 248, 285, 352, 398, 410, 422, 468.)* The other national centres had sometimes difficulties in obeying the provisions of the Conventions.

International exchange of publications extended and developed also in the countries that had not signed the Brussels Conventions. For example The International Exchange Service of France, which was founded as early as in 1877, attended to all the tasks prescribed in the Convention. Some Latin American countries founded exchange centres in connection with their national libraries at the end of the 19th century.[3] In Finland, which at that time was a part of Russia, a national exchange centre was founded by the scientific societies in 1899. The British Museum was active in extending exchanges. The University library of Uppsala, which had a long tradition in exchange, increased its connections many-fold. The successful development in Sweden was partly due to the co-operation between the University and scientific societies, partly to the privilege of exemption from postage since 1885 and partly to the sufficient library staff. (Handbook *1956, 262-263, 266, 283, 289, 29 ; Harris 1998, 427-428 ; Bring 1929, 133-135.)* As much as international agreements, exchange of publications needed support from the local government and sufficient resources to flourish.

Many new conventions concerning the exchange of publications were signed before World War I. Most of these were bilateral cultural agreements between governments and they dealt mostly with official publications. For instance Belgium signed in the 1890's agreements with France, Netherlands, Luxembourg and Hungary. (Manuel 1950, 58.) The first regional multilateral convention was the Inter-American Convention signed in Mexico in 1902. It was signed by the Argentine Republic, Bolivia, Chile, Colombia, Costa Rica, the Dominican Republic, Ecuador, El Salvador, Guatemala, Haiti, Honduras, Mexico, Nicaragua, Paraguay, Peru, United States of America and Uruguay. The aim of this convention was to further mutual understanding and closer integration between the countries of the American continent and it was not mutually exclusive with the Brussels Conventions. However, these conventions differed in some points from each

---

[3] The Latin American countries were: Colombia (1868), El Salvador (1870), Chile (1874).

other. For example, according to the Inter-American Convention, the material was to be sent through diplomatic channels, not through exchange centres. Also the definition of official publications was more comprehensive in the Inter-American Convention including also geographical maps, plans and works of all kind subsidized by the respective signatory governments. (*Handbook* 1964, 55-56, 69-71.)

The exchange of institutes, scientific associations, museums and libraries was usually still based on informal correspondence between these institutions. However, when exchange relations increased, printed form letters or postal cards gained ground to the exclusion of personal letters and intercourse. Before the World Wars libraries were still quite eager to get as many exchange partners as possible, whether the publications received were relevant for them or not. The age of the information flow had not yet really begun. Exchange of publications was very important for libraries, because purchase of periodicals was expensive due to the currency fluctuations. (Gibson 1982, 157-158.)

The active phase of the international exchange of publications launched by the Brussels Conventions and other agreements was interrupted by the World War I. The war had severe effects on European libraries: difficulties in financing and staffing, destruction of libraries and the rupture of diplomatic relations. After the war, the political situation in Europe was altered. Germany was militarily weak and in economic bankruptcy, the revolutions in Russia had created the first socialist state in the world and new independent states had emerged from the Russian, German or Austrian territories. (Harris 1984, 192.)

Soon after the Peace of Versailles (1919) a new phase of development began in the exchange of the publications. Rumania and the new independent states Czechoslovakia, Poland, Hungary, Latvia and the City of Danzig adhered to the Brussels Conventions in the 1920's. Also China and Egypt ratified the Conventions thus extending the area outside the European and American continents. Bilateral agreements were still important. For example Germany, which had not adhered to the Brussels Conventions, was active in concluding bilateral agreements in the 1920's and 1930's. (Manuel 1950, 56-85 ; Handbook 1956, 56-57.)

A new Inter-American Convention was concluded in Buenos Aires in December 1936. Signatories were the same states as in the previous Convention of 1902, added with Brazil, Cuba, Panama, and Venezuela. This new Convention emphasized the role of national or official libraries as beneficiaries of exchange. (Handbook 1956, 56-59 ; Handbook 1964, 55-56.) Despite the Pan-American spirit of these Inter-American conventions, the development of libraries in Latin America was very uneven. The countries with poor economic resources or unstable governments were not able to found modern libraries. However, support of the international co-operation encouraged library development in Latin America. (Harris 1984, 275-281.)

Despite the international goodwill towards exchange, many improvements were still needed. Postal charges and customs fees made despatch of publications expensive for libraries. The Commission of Intellectual Co-operation of the League of Nations tried many times, but unfortunately without success, to secure international postal franchise on exchange material. Individual countries could, however, negotiate bilateral agreements of free postage. For example The Smithsonian Institution had succeeded acquiring the right of free postage to a large number of countries on the American continents. (Manuel 1950, 100-101, 132-134.)

The major concern for The International Federation of Library Associations, IFLA, founded in 1927, was the expanding exchange of university dissertations. The problem had a long history beginning from the dissolution of *Akademischer Tauschverein* in the end of the 19th century. IFLA's activity was, however, sporadic and unsystematic and not much was effectuated. *(Gombocz 1974, 9-10, 16).*

Even though the exchange of publications expanded and many national centres were founded, it became evident that Brussels Conventions never had in fact become the world-wide conventions which they were designed to be. They also contained some very weak spots, which the League of Nations through its International Commission of Intellectual Co-operation tried to remedy. These efforts were interrupted by the World War II. (Handbook 1956, 56.)

World War II and the Period of the Reconstruction

The destruction left by the World War II was much more severe than the damages of World War I. The collections of national libraries were usually evacuated, but bombardments destroyed or heavily damaged many university libraries in France, Germany, the Soviet Union and smaller countries. (Harris 1984, 197-200 ; Harris 1998, 553-563.) Though it sounds peculiar, the war had also some positive consequences for libraries. Modern information service developed as rapid conveying of information was necessary for the war technology, military operations and espionage. (Spence-Richards 1997, 174-178.)

The war could destroy the library buildings and collections, but not the idea of scientific or cultural exchange. All intellectual contacts between the Axis states and the Allies were suspended during the war, but contacts were maintained inside these blocs. Naturally, the quantity of official as well as scientific publications decreased and transports were risky. Nevertheless, exchange services stored up publications to dispatch to their correspondents after the war. The havocs caused by bombardments made it necessary to search international co-operation. Hence the development of exchange activity was one aim in the Conference of Allied Ministers of Education, held in London in 1943. This meeting was also an important step towards the establishment of Unesco. (UNESCO ; *Manuel* 1950, 102.)

The Unesco Preparatory Commission met in London in 1945 and examined, among other things, the question of exchange of publications. In the First General Conference of Unesco in 1947, Unesco's Clearing House for publications was established. It was meant to be the main centre for the promotion of all direct exchanges between institutions throughout the world. This was only the beginning of activities. The Committee of Experts convened in 1948 and gave Unesco recommendations concerning the exchange of publications: it should promote the bilateral inter-governmental agreements and make a draft for them; it should also promote the establishment and development of the exchange bureaux; bibliographies should be published of official publications, university publications and publications issued by learned societies, scientific institutes, etc.; there should be lists of existing national exchange bureaux and of institutions able to supply information on exchange possibilities, and reports on the activities of exchange bureaux. (*Manuel* 1950, 102-105 ; *Handbook* 1956, 11.)

Hence Unesco had an immense field of work and it could progress only bit by bit. *The Unesco Bulletin for Libraries*, launched in 1947, was one answer to the long-lasting problem – lack of information on exchange possibilities. The bulletin included sections "Publications wanted", "Exchange" and "Free Distribution", so that libraries could find new exchange partners or duplicates available. (*Handbook* 1956, 47)

The next step of Unesco was a questionnaire to all exchange bureaux asking for detailed information on existing exchange arrangements and suggestions for their improvement. M*anuel des échanges internationaux de publications - Handbook on the International Exchange of Publications* published in 1950 was based upon the replies to this questionnaire. It included a brief history of exchange; a list of exchange contracts; a list of exchange bureaux; a bibliography of lists of official publications and a classified list of institutions willing to exchange publications. The text was in French and in English. This kind of handbook was really needed, but unfortunately, it remained a torso, because the answers returned for the questionnaire in 1948-1949 were very incomplete. Many countries had not yet recovered from the effects of the war, exchange relations had not yet been restored and thus there was nothing to report. In some countries even the postal services did not function. Conditions everywhere were, however, undergoing a radical change. Hence, already in 1951 Unesco considered issuing a supplement to the Handbook and at the Seventh Session of the General Conference in 1952 it was resolved to issue a completely new edition. The second edition of the Handbook included more detailed and up to date information of the exchange centres arranged by countries. New sections were added on the international organizations. This time the text was in English, French and Spanish. (*Manuel* 1950, 93-94 ; *Handbook* 1956, 11-13.)

Unesco's efforts were not restricted to conveying of information. It also made efforts to establish exchange centres in Canada, Cuba, Egypt, Greece, India, Indonesia, Iran, Mexico, Philippines, Sweden and Turkey. Some of these countries needed help because they did not have enough resources, but some of them just did not prefer the centralized exchanges. For instance, the University library of

Uppsala, which was among the most active exchanging libraries in the world, preferred single-minded individual exchanges, because they considered that exchange centres would not save expenses and transmission would be slower. (*Handbook* 1956, 40, 47.)

One of the long-term problems of the exchange of publications was the question of transmission. The exchange bureaux tended to use the cheapest way of transport, which was normally not the most rapid way. Encouraged by the member states of Unesco, Universal Postal Union admitted in 1952 each administration the option of conceding a 50 per cent reduction on the rate for all printed matter, whoever the sender. Many countries adopted this reduction, but with certain limitations, for example excluding periodicals. Not only postage, but also custom fees oppressed the exchange. The senders of exchange material often failed to fill the customs forms correctly, with the result that the addressee had to pay higher customs duties than necessary. *The Agreement on the Importation of Educational, Scientific and Cultural Materials* was signed in 1950 to solve this problem. It obliged the contracting states not to apply customs duties or other charges on the import of educational, scientific and cultural material. However, up to the year 1955 only 18 states had ratified the Agreement. (*Manuel* 1950, 132, 135 ; *Handbook* 1956, 61-70.)

Unesco was not alone in developing the exchange of publications and helping the libraries. IFLA continued its efforts to develop the exchange of dissertations when it met the first time after the war in 1947. However, by the abolition of compulsory printing of dissertations in Germany and France the overflow of dissertations was eased. Microfilms were expected to be a solution to the problem of distributing theses. (Gombocz 1974, 10-11, 16.) United States Book Exchange Inc. (USBE) was established in 1948 to serve libraries by receiving and distributing duplicate scientific books and periodicals (*Handbook* 1956, 29, 39-40, 469).

Many international organizations took an optimistic attitude towards the exchange of publications and had programmes for developing it. The second edition of the Handbook (1956) mentioned 94 international organizations interested in exchange including library associations, professional organizations, religious and scientific associations. Organization for European Economic Co-operation (OEEC) had a large scheme for the exchange of publications, which came into being in 1950. The idea behind this *OEEC Documents Exchange Scheme* was that in order for the various industrial fields to be on the same level in each country, research findings should be available to all. Trade and industry should also have access to the "grey literature" research publications, which were not for sale. However, this scheme did not prove satisfactory and in a meeting of technical information officers in 1959 it was abolished. (*Handbook* 1956, 44-48, 207-214 ; *Handbook* 1964, 55.)

New multilateral agreements were made between the Arab League in 1945 and the Spanish speaking countries in 1953. They were actually cultural agreements, which contained clauses referring to the exchange of publications. However a few years later, The Arab States Conference on the Exchange of Publications (Damascus, 1957) gave more exact instructions to the exchange recommending that deposit and exchange centres should be housed at national libraries, which should also publish national bibliographies and lists of publications offered for exchange. (*Handbook* 1956, 50-52 ; *Handbook* 1964, 87.)

The end of the 1940's and the beginning of the 1950's was the period of growing optimism towards international co-operation. The chairman of IFLA, A. Kessen, could declare already in 1953 that "Exchange relations disturbed by the war have been restored". (Gombocz 1974, 11.) Restoration of exchange relations was nevertheless not enough to satisfy the information needs of universities or scientific institutes in industrialized countries. Modern libraries needed scientific information much more rapidly and efficiently than earlier. Neither had Unesco's idea of access to publications for people of all countries, yet come true. Many problems still lay unsolved.

Unesco Conventions

The weak spots of the Brussels Conventions had already been noticed before the World War II, but improvements were not possible before the 1950's. In the post war situation three groups of insufficiencies were discussed. The old definition of an official publication caused trouble. All countries were partners with equal rights and this meant that big countries had to give away incomparably more than they received, while the smaller countries received an overflow of publications, which was difficult to handle. The other problem was the strict restriction of exchange bureaux taking initiative in exchanges. The conditions of transport and customs had also to be reviewed. Especially France, Italy and the socialist countries encouraged Unesco to plan a new convention. The representative of IFLA proposed a revision of the Brussels Convention. In the meeting of experts in 1956 a draft convention was prepared by the Unesco Secretariat and sent out to member states for comments. In the tenth session of Unesco's General Conference in Paris in 1958 two new conventions were approved and signed.

The Convention 1, *Convention Concerning the International Exchange of Publications,* was something absolutely new, because it dealt also with scientific and literary publications. All previous multilateral conventions had been concerned only with official publications, except the Inter-American Convention of 1936, which was regional in character. The provisions of this convention were quite flexible in order to be applicable also in countries that had strict control over their learned institutions. The Convention 2, *Convention Concerning the Exchange of Official Publications and Government Documents Between States,* aimed to avoid the weak points of the Brussels Conventions. Thus the definition of official publications was not very specific and the contracting states had the right to determine the official publications and government documents, which were to be used as exchange material. Bilateral agreements were accepted for the purpose of implementing the Convention.

In both Conventions an authority apart from an exchange centre could take responsibility of transmitting the publications or supplying information. Transmission could also be made directly between the exchange partners. Both Conventions recommended that contracting states should try to ensure that the exchange authorities had the most favourable transport rates and that each state should grant exemption from customs duties for imported and exported material. A new feature was an obligation to send annual reports to Unesco, which promised to publish this information. Unesco was presumed to give technical assistance with any problems arising from the application of these conventions. (*Handbook* 1978, 63-65 ; Gombocz 1974, 12 ; *Handbook* 1964, 55-56, 62-69 ; Gombocz 1973, 38.)

Both Conventions came into force in 1961. By 1976 thirty-nine states had accepted or ratified them. Among the ratifying countries were the Soviet Union, the United States and many major European countries, but very few developing countries. In the Conference of the International Exchange of Publications in Europe (Budapest 1960) the participants were hopeful, that if only a sufficient number of states would ratify these Conventions, they would become an efficient tool in developing the exchange of publications. (Cox 1962, 80 ; *Conférence 1962,* 234.) This optimism faded gradually. The Conventions have been criticized for being so vague that a country can endorse their principles without necessarily taking serious action. According to the inquiry of István Gombocz in 1971, no remarkable development had happened during ten years in the exchange of the signatories. For Unesco, the Conventions have, however, been an important tool in carrying out library and information programs. Like the Brussels Conventions, the Unesco Conventions have effected also to the countries that had not signed them providing general principles that any library can use in establishing exchange relations. *(Handbook* 1978, 66 ; Gibson 1982, 156 ; Popov 1981, 37.)

New edition of the Handbook was needed in the beginning of the 1960's, not only to introduce the Unesco Conventions, but also to cover an increased production of scientific publications. The third edition of the Handbook was published in 1964. It was an enlarged up-to-date version of the second edition with the text in English, French, Spanish and Russian. *(Handbook* 1964, 11.)

The overflow of publications became gradually a more and more remarkable problem for libraries and it had an effect on the exchange of publications. Many institutions diminished the amount of their exchange partners or abandoned the exchange alltogether. (Gibson 1982, 158-159.) In the fourth edition of the Handbook, published in 1978, the editors had already found it impossible to keep the lists of publications available in exchange completely up to date. Instead the fourth edition gave more accurate directions, how to handle and organize exchange, for example recommending different card indexes and standardized request forms. *(Handbook* 1978, [5-6], 38-44.)

Exchange of Publications in the Divided World

In the second edition of the *Handbook* El Salvador, Spain, Luxembourg and the Netherlands had made known that they did not make exchanges with the Soviet Union or countries of Eastern Europe.[4] The Iron Curtain had descended on Europe and this had an influence on the exchange of publications. The poles of the divided world, the United States and the Soviet Union, did not, however, limit their exchange within their own blocs. Actually, both of them were quite active in promoting international exchange. The Budapest Conference of the International Exchange of Publications, in 1960, was organised partly to bring the countries of Eastern and Western Europe closer to each other. Most of the discussions were, nonetheless, quite cautious. Exchange of publications was seen as an important way of maintaining peaceful international co-operation by all parties. G. A. Hamel, the Head of the International Exchange Bureau of the Royal Library of the Netherlands, stated in his working paper, that the most formidable obstacles to the exchange are the political ideas, which prevent free access of material to some countries. The representatives from Bulgaria and Soviet Union complained that publications from their countries had been sent back. This question was not, however, discussed in detail. The resolutions of this conference were quite diplomatic: the freedom of science was emphasized and it was recommended that the co-ordination of national exchange centres should never become a matter of control or centralization. Free initiative should be left to the exchanging bodies. The Conference also recommended visits and meetings of the exchange professionals. *(Handbook* 1956, 283, 285, 377, 381, 442 ; *Conférence* 1962, 98-199, 200-202, 232, 239-241.)

The world was divided not only to East and West, but also to North and South. In the first and second edition of the Handbook many African and Asian countries were still mentioned in connection with their mother countries. In the third and the fourth editions the situation had changed: the colonies had become independent and aimed to found exchange relations of their own.[5] These former colonies had, however, many difficulties in finding their place in the international network of libraries, and the former mother countries had problems as well in orientating to this new situation. The mother countries had the right to legal deposit in the colonial period, but now they were expected to be the giving party.

Many developing countries could not afford to establish exchange centres. Furthermore, lack of professional staff, language barriers and difficulties in transportation and communication complicated the exchange of publications. (*Handbook* 1978, 117, 138, 149 ; Vanwijngaerden 1981, 48-51.) For their exchange

[4] El Salvador excluded also Asian countries; Luxembourg Scandinavian countries; Netherlands China and Indonesia.

[5] For example Congo, Malawi and Senegal had no organized exchange centres, but the libraries and universities were announced to be willing to exchange publications.

partners it was difficult to find institutions interested in exchange, if no national exchange centre existed. The other difficulty was maintaining the balance of exchange, because the developing countries could not offer a sufficient number of publications.

These problems were already mentioned in the Budapest Conference of 1961. In the Vienna Congress of 1972 the situation of the developing countries was discussed more thoroughly. The resolutions of the Vienna conference proposed that European exchange partners should take a more generous attitude, without primary considerations of commercial value. IFLA was asked to inform its member-associations that they might request technical aid from Unesco for the establishment of national exchange centres. *(Conférence* 1962, 237 ; *International Exchange 1973,* Discussions, p. 109-113, Resolutions, p. 133-134.)

Language barriers were one of the major problems, which were discussed in the IFLA meetings as well as in the conferences of Budapest and Vienna. The chairman of IFLA, A. Kessen, turned to the Scandinavian countries, Japan, Turkey, Egypt and Soviet Union suggesting that all dissertations not written in a well-known language should have a summary in one of the better-known languages such as English, French or German. This proposal was quite largely approved in Japan and the Scandinavian countries. The Leningrad Academy introduced summaries from 1956 on. (Gombocz 1974, 11-12, 16 ; *Conférence* 1962, 237 ; *International Exchange* 1973. Discussions, p. 109-113, Resolutions, p. 132.)

The problems of libraries in the 1960's and the 1970's were very different from Oldenburg's or Vattemare's days. Scientists did not have a common language, as they had in the 17th century, when everybody understood Latin. Publications had to be delivered rapidly, because researchers could not wait for months for an interesting article. Commercial distribution of publications had become an efficient and rapid way of acquiring scientific literature, usually written in English. Open access to information had become an ideal, at least in the western countries. The "baby boomers generation" started studies in universities expecting to find research material they needed in the university library.

The exchange of publications has been criticised from many standpoints in the 1960's and the 1970's. It has been said, that the exchange disturbs the commercial distribution of publications, because scientific institutions must deliver the exchange copies, which otherwise could be sold. From the viewpoint of libraries, the exchange takes up more work than the purchasing of material. Some research has shown that material acquired by exchange is not used as much as purchased material. (Södergård 1992, 11-12.) These criticisms are partially justified, but they do not hold true for all situations. For instance, the exchange is not a threat to commercial distribution of those publications that do not sell very well or are not distributed commercially at all. For many countries the lack of foreign currency still limits library purchases to a minimum and thus the exchange is the only way to acquire foreign literature. The exchange of publications is also a natural part of international cultural and scientific co-operation. According to inquiries of the IFLA, the view to the exchange of publications was still quite optimistic at the end of the 1970's. Many libraries planned to enlarge their exchange and six libraries were so far-sighed that they planned to computerize the exchange procedures. (Genzel 1981, 11-15; Kanevsky 1981, 23-26.)

The optimism had its ground. Exchange of publications had for over three hundred years been an important way to disseminate information as well as to acquire publications. It had developed side by side with libraries, scientific publishing and diplomatic and cultural relations. In the beginning the innovative personalities created new methods in establishing exchange relations and making exchange run smoothly in libraries and scientific societies, academies and museums. From the 19th century on the national centres and international multilateral conventions had become more important. After the World War II especially the activities of Unesco and IFLA had managed to overcome many difficulties. The development of the exchange of publications is, however, always dependent on persons devoted to this activity. No international organisation or a convention can be created without the enthusiasts who believe in the ideal of free access to information for all people of the world.

References

Armbruster, Carol (1997) The Origins of International Literary Exchanges : Alexandre Vattemare's Adventures in America, *Biblion : the Bulletin of The New York Public Library,* Westport (CT), Vol. 5, nr 2, Spring; 1997.

Berninger, Ernst H. (1997) Die Bibliothek des Deutschen Museums als Archiv und Forschungsbibliothek für die Geschichte polytechnischer Kultur, in *Technische und naturwissenschaftliche Bibliotheken in ihrer historischen Entwicklung und Bedeutung für die Forschung,* Wiesbaden.

Bring, Samuel E. (1929) The International Exchange Relations of the University Library of Uppsala, Sweden, and their Organization, *Nordisk tidskrift för bok- och biblioteksväsen,* Uppsala, Årg. 16.

Broman, Thomas (2000) Periodical literature, in *Books and the Sciences in History,* ed. Marina Frasca-Spada and Nick Jardine, Cambridge.

Clark, William (2000) On the bureaucratic plots of the research library, in *Books and the Sciences in History,* ed. Marina Frasca-Spada and Nick Jardine, Cambridge.

Conférence sur les échanges internationaux de publications en Europe. (1962) *Budapest, 13-19 septembre 1960,* Budapest.

Cox, W. (1962) Present Trends and Future Possibilities in the International Exchange of Publications, in *Conférence sur les échanges internationaux de publications en Europe. Budapest, 13-19 septembre 1960,* Budapest, 1962.

Exchange of Publications (1972), in *Encyclopaedia of Library and Information Sciences,* ed. Allen Kent and Harold Lancour, New York, Vol. 8.

Genzel, Peter (1981) Universal Availability of Publications and the Section on the Exchange of Publications. Report on its activities 1970-1979. In *Studies in the International Exhange of Publications*, ed. by Peter Genzel, *IFLA Publications, 18,* München.

Gibson, Sarah S. (1982) Scientific Societies and Exchange : a Facet of the History of Scientific Communication, *Journal of Library History,* Austin (MN), Vol. 17, No 2, Spring 1982.

Goldgar, Anne (1995) *Impolite Learning : Conduct and Community in the Republic of Letters 1680-1750*, New Haven.

Gombocz, István (1973) The Unesco Conventions of 1958, in *The International Exchange of Publications : Proceedings of the European Conference held in Vienna from 24-29 April 1972*, Ed. Maria J. Schiltman, Pullach/München,.

Gombocz, István (1974) The forty Years of the Committee on the Exchange of Publications, edited by Maria J. Schiltman, *IFLA Journal*, (Pullach-München),Vol. I.

Handbook on the International Exchange of Publications (1956). 2nd ed., Ed. Gisela von Busse, Paris.

Handbook on the International Exchange of Publications (1964) 3rd ed., Ed. Gisela von Busse, Paris.

Handbook on the International Exchange of Publications (1978) 4th ed. Ed. Frans Vanwijngaerden, Paris.

Harris, Michael H. (1984) *History of Libraries in the Western World,* Metuchen (NJ).

Harris, P. R. (1998) *A History of the British Museum Library 1753-1973*, London.

The International Exchange of Publications : Proceedings of the European Conference held in Vienna from 24-29 April 1972, ed. Maria J. Schiltman, München, 1973.

Jörgensen, Arne (1930) *Universitetsbiblioteket i Helsingfors 1827-1848*, Publications of the University Library at Helsinki, 44, Helsinki.

Kanevsky, P. B. (1981) International Book Exchange of the National Libraries in the Late 70's, in *Studies in the International Exhange of Publications*, ed. by Peter Genzel, *IFLA Publications, 18,* München.

Kleinert, Andreas (1997) Vom Buch der Natur zum Druckerzeugnis : aus der Geschichte der naturwissenschaftlichen Fachliteratur, in *Technische und naturwissenschaftliche Bibliotheken in ihrer historischen Entwicklung und Bedeutung für die Forschung,* Wiesbaden.

Manten, A. A. (1980) Development of European Scientific Journal Publishing before 1850, in *Development of Science Publishing in Europe*, Ed. by A. J. Meadows, Amsterdam.

Manuel des échanges internationaux de publications, (1950) préparé avec le concours de Mlle J. L. Dargent, Paris.

Popov, Vladimir (1981) The Multilateral Unesco Exchange Conventions (1958) and their International Significance, in *Studies in the International Exchange of Publications*, ed. Peter Genzel, Ifla Publications, 18, München.

Spence-Richards, Pamela (1997) Der Einfluss der US-Regierung auf technische Bibliotheken im 20. Jahrhundert, in *Technische und naturwissenschaftliche Bibliotheken in ihrer historischen Entwicklung und Bedeutung für die Forschung*, Wiesbaden.

Södergård, Pia (1992) *Bytesverksamhets betydelse för de vetenskapliga biblioteken i Finland : en förstudie*, Meddelanden från Ekonomisk-statsvetenskapliga fakulteten vid Åbo Akademi. Institutionen för biblioteksvetenskap och informatik. Ser. A: 362, Åbo.

UNESCO 1945-2000 : a Fact Sheet", The web site of Unesco, http://www.unesco.org/general/eng/about/history/back.shtml, [6 June 2003].

Vallinkoski, Jorma (1948) *The History of the University Library at Turku, 1*, Publications of the University Library at Helsinki, 21, Helsinki.

Vallinkoski, Jorma (1975) *The History of the University Library at Turku, 2, 1722-1772*, Publications of the University Library at Helsinki, 37, Helsinki.

Vanwijngaerden, Frans (1981) Exchange of Publications with Developing Countries, in *Studies in the International Exchange of Publications*, ed. Peter Genzel, Ifla Publications, 18, München.

Wyatt, Robert , 1997. Die Entwicklung wissenschaftlicher und technischer Bibliotheken in Grossbritannien, in *Technische und naturwissenschaftliche Bibliotheken in ihrer historischen Entwicklung und Bedeutung für die Forschung*, Wiesbaden.

Zur Geschichte des akademischen Tauschvereins (1885) *Centralblatt für Bibliothekswesen*, Leipzig, 2. Jahrg., 1885.

THE INTERNATIONAL EXCHANGE OF PUBLICATIONS AS AN ACQUISITION METHOD IN ACADEMIC AND SPECIAL LIBRARIES : THE CASE OF FINLAND REVISITED

Pia Södergård

Åbo Akademi University Library

Finland

The international exchange of publications is an activity with roots in the 17th century, which has stood up to hardships through times (Gibson 1982, 144-163). As an acquisition method in libraries it has excelled in unstable areas where the book trade has been unorganized and political conditions volatile (Barker 1986, 63-73 ; Yu 1981, 333-338). When the objectives and methods of managing collections have changed and developed, the merits and demerits of the exchange of publications as an acquisitions method has over and over again been discussed in the literature. Increased commercialization and emphasis on resource sharing were two challenges of the past toward the activity (Herslow Fex 1983, 7-11 ; Kovacic 1980, 155-163). Nowadays, new technology deeply affects libraries and puts pressure on many different levels but new economic models and new user perspectives also have an impact on how libraries function and develop.

As regards the impact of new technology, the following trends have been emphasized: Firstly, the parts played by authors, publishers, libraries and other information service providers are changing, and the boundaries, which have demarcated the roles of these players, have become blurred (Dorner 2000, 15-44). Secondly, each player in the 'information business', not least the libraries, has been forced to consider their role anew as their very existence have been questioned (Brophy 2001, 219). Whose core business is about to disappear? Who will become superfluous? What are the primary roles which libraries are going to fill in the future? Thirdly, for the individual library new technology offers a potential to save money, shelf space, and library staff time while simultaneously increasing the rapid provision of information to patrons (Nisonger 1997, 29-57). Daniel G. Dorner (2000, 15-44) emphasizes that while all types of libraries are being transformed, the changes are occurring at various speeds and to varying degrees in different types of libraries and for different types of library operations.

In Finland, the development of electronic dissemination of knowledge was furthered by two measures taken by the Ministry of Education. In 1995 new recommendations were issued for the universities with regard to the reform of the methods of publishing doctoral dissertations. The recommendations stated, among other things, that dissertations which are being defended for the doctoral degree should not be printed and that libraries should abandon the use of dissertations in exchange of literature because the character of the dissertations will change (Häkli & Lehtinen 2000, 14-25). By the end of the century, Finnish university libraries had become very active in the process of organizing the electronic publishing process in their respective universities (Sonkkila 2000, 99-100). In 1997, the National Electronic Library Programme, FinElib was launched to support higher education, learning and research in Finland. The basic goals of the programme are to increase the quantity of electronic information available to users, to improve information retrieval from the Internet and to develop a graphical user interface. The programme is based on shared financing between the university libraries combined with a subsidy from the National Library (Hormia-Poutanen 2001, 17-21). At the moment there are about 6000 full-text online journals, 90 reference databases, dictionaries, and reference books available to users (FinElib 2003).

The influence of new economic models can be quite fundamental. In the 90's, the state introduced a new budgetary system, which was based on the organization's planned results instead of on its budgets in the previous years. As an incentive the organizations were promised that with the new system they would have more freedom in using the resources allocated in the budget (Järvenpää 1998, 22-24). When universities implemented the new result-oriented budgetary system, academic libraries were no longer regarded by their universities as part of a nationwide, joint library system. They were expected to be supportive of operations and functions in their own frame-organizations. The tighter the budget, the more libraries were required to provide for the current needs of their university (Häkli 1996). In the early 1990's, due to economic recession, libraries faced a time of diminishing budgets and increased demand on service.

In the year 1989, before the above-described developments took place in Finland, data on the exchange programmes in Finnish academic and special libraries were collected by means of a questionnaire for an investigation about the scope, objectives and administration of the activity (Södergård 1992 ; Södergård 1993, 104-109).14 years later, in the winter of 2003, a small-scale follow-up study was conducted by means of a telephone interview in order to find out the current status of the activity. This article presents and compares these two studies in broad outline.

1989 — a Flourishing and Stable Activity

In 1989, all Finnish academic and special libraries which appeared on the list of libraries contributing to official statistics in Finland received a questionnaire about exchange activities. The official statistics had shown that slightly less than half of the acquired material to these libraries was purchased (Tieteellisten... 1988, 67). 51 libraries responded which yielded a reply rate of 89%.

At that time the exchange of publications was a flourishing activity in Finland. 80% of the responding libraries had some form of exchange programme. The total sum of agreements for the participating libraries was 8,837, varying in size from five to 595 agreements in the individual libraries. A categorization of libraries according to the size of their exchange programmes showed an equal distribution for libraries having a small-scale, a medium-scale and a large-scale programme. About a third of the libraries had less than 100 agreements, a third between 100 and 300, and a third over 300 agreements. No differences between academic and special libraries could be observed. Geographically the scope of the exchange programmes was narrow. An average of 83.4 % of the exchange partners were from Europe and 9.9 % from North America.

The data gave also evidence for a very stable activity. In a preceding ten-year period, only two libraries had cut down the size of their programmes and no plans of elimination or dramatic curtailment of programmes were reported. The libraries either endeavoured to maintain the amount of agreements they had established or planned to slightly increase the scope of their programmes in order to include new countries or new subject areas. During the preceding twenty years, eight libraries had initiated new exchange programmes as soon as they were given the opportunity to do so.

The libraries were asked why they maintain exchange programmes and their answers could be grouped into six broad categories. The answers of most of the libraries could be related to two or more categories but all six categories were powerful enough to appear as a single objective as well:

A) Convenience (17.7 %). Exchange of publications is a convenient way to acquire material. It is easy to monitor exchange activities. The reliability and regularity of exchanges as an acquisition method were stressed.

B) Financial saving (34.1 %). The exchange programme enables the library to acquire more material than the materials budget would allow. It is a way of receiving material "free of charge".

C) The material (63.4 %). The programme is maintained in order to acquire material of value and importance to the collection. Material that is not available through commercial channels, and certain types of material such as academic dissertations received special mention.

D) PR (51.2 %). The exchange programme is a PR-tool for the researchers of the frame organizations of the libraries. The publications of the frame organization are distributed and made known through the exchange programme. The international exchange of publications is regarded as a form of scientific communication. Through exchange agreements international contacts are established and maintained.

E) External motives (24.4%). For reasons not originating in the needs of the library. The exchange programme is acknowledged to be of vital importance to the exchange partners of the library especially to those in countries with a shortage of foreign currency. Sense of tradition and free access to the publications of the frame organization were also mentioned.

F) Other reasons (7.3 %). This category included unspecified answers of the type "to fulfil a need".

A difference between academic and special libraries was apparent. The amount of libraries which mentioned PR was larger in the group of academic libraries. Special libraries stressed financial saving and convenience more strongly.

The value of the material was the most powerful category of them all. At that time however, the material received through exchange relations was also very much in question. Firstly, the problem of receiving unwanted material was of concern. The receipt of not needed material, which is expected in acquisition forms based on mechanical decisions criteria, was recognized by slightly over half of the libraries who reported that they received material that was not added to the collection. On an average about 15 % of the material was superfluous. An average of 12 % was added to the collection without any processing and an average of slightly over 70 % was catalogued. Secondly, the actual value of the exchange material was an issue. Does acquiring material of value yield valuable collections? Earlier investigations showed that collection development librarians felt that the material received through exchanges was of crucial importance to the development of collections (Yu 1981, 333-338 ; Deal 1989, 199-207). On the other hand, there were distressing results from case studies, which proved that the exchange material was notoriously seldom used and could thus be of very little value for the library (Herslow Fex 1983, 7-11 ; Kovacic 1980,155-163 ; Fjällbrant 1984, 77-86). It was put forward that the approach to collection development, as far as exchanges are concerned, seemed to be publication based rather than user based (Södergård 1992). By considering the four different roles of collection described by Michael Buckland (1991, 225) a more refined picture emerged: User studies only measure how well a collection is fulfilling its *dispensing role* i.e. to provide convenient physical access to copies. Collection development librarians probably take into account the other roles of the collection as well. The *preservation role* is fulfilled by collecting and preserving documents, which would otherwise be lost. *The identifying or bibliographic role* shows itself in libraries where collected items can be scanned directly and gives a picture of what material exists. The *symbolic role* is manifested by the feeling that large collections, particularly of special materials, bring status and prestige.

A European survey showed that Finnish collections did not fulfill their dispensing role well on a national level. Finland was very dependent on literature and information resources from abroad. In comparison to other countries it had the second largest number of requests sent to libraries outside the country (Cornish 1990, 39-52). The reasons for this were explored by different methods but no weaknesses in national literature and information provision could be detected. A comparison with Sweden, who was a large supplier to Finland, showed that the yearly resources spent on acquisitions was about the same. In Finland, however, the amount was scattered on 42 units and in Sweden on only 26. The large number of units in the Finnish university network inevitably led to spending resources on maintaining many overlapping core collections which meant that the material most in demand was amply available but less used material was scarce. Moreover, a citation analysis of Finnish and Swedish doctoral dissertations provided evidence supporting the assumption that there are national features in research literature use in Finland. Finnish researchers used much more literature in the literature intensive subjects of social studies and humanities than their Swedish colleagues and in medicine, Finnish researchers' use of literature was scattered over a wider range of sources (Södergård 1997).

2003 — a Decreasing Activity under Observation

In the winter of 2003 a small-scale telephone interview[1] was conducted with Finnish academic and special libraries in order to explore the current status of the international exchange of publications as an acquisition method. 53 libraries in all were contacted and 50 libraries participated in the survey. It was basically the same libraries as in the earlier survey although there were libraries, which had merged, a few libraries had changed their names and some had moved during this period.

70 % of the libraries that participated in the survey have some form of exchange activity, which is a decrease by 10% compared to 1988. An interesting difference between academic and special libraries emerged since 83 % of the special libraries have an exchange program whereas only 58 % of the academic libraries maintain programmes. It was however not possible in these telephone interviews to go deeper into why libraries had terminated their programmes. Many years had passed and often the persons who acted as informants on behalf of the library did not have the historic perspective required to answer such a question. One academic library which had recently shut down its exchange programme referred to a change in collection development policy. After having decided to abandon the idea of complete collections and to refrain from acquiring series in full, a natural consequence was to reflect on the exchange programme. The fact that the publications of the frame organization were made available on the net contributed to the conclusion of the programme. Another library reported that the programme was terminated in the 90's when they were expected to pay for the publications they had been using for exchange purposes.

The size of the exchange programmes had considerably diminished. The total sum of agreements for the libraries was now only 3,419 varying in size from four to 455 agreements in the individual libraries. The decline from 8,837 to 3,419 seems striking but is somewhat alleviated by the fact that a substantial number of agreements have been transferred to The Exchange Centre for Scientific Litera-

[1] Ms Birgitta Eriksson-Katajainen's assistance in collecting the material is gratefully acknowledged.

ture in connection to the merging of libraries. Furthermore, the tidying up of records from inactive exchange relations, which was reported by libraries, is also worth mentioning. When looking at the individual libraries, the majority of the libraries now had a small-scale programme, 62.9 % of the libraries had fewer than 100 agreements. The group of libraries with a medium-scale programme was 29.4 %. Only 3 libraries (8.8 %) were found with over 300 agreements compared to 14 in the earlier survey and only two libraries reported an increase in agreements.

The scope of the programmes was not investigated but it seems inevitable that the curtailment of programmes has narrowed their scope. One library reported having cancelled all domestic exchanges, another library does not have any foreign relationships left. The language of the publications was also mentioned as a criterion for pruning.

The libraries were asked about the current motivations of the programmes but the difference in data collecting method makes it difficult to compare the answers to those in the earlier survey. The fact that the objectives mentioned now were fewer than in 1989 could indicate that the activity can no longer be as easily motivated but it could also just mean that it is easier to answer such a question in writing at leisure than on the telephone. For the academic libraries, the strongest motivation still was to acquire material of value and importance to the collection. The needs of the exchange partners were not mentioned at all and only a third of the libraries mentioned PR-aspects. The special libraries, which as a whole gave more motivations, emphasized the convenience of the method as strongly as the material acquired. The needs of the exchange partners were mentioned by a fourth of the libraries and PR-aspects were emphasized more than financial savings.

Slightly over half of the academic libraries reported that the material available for their exchange programme had also been made available on the net compared to only three special libraries. In most cases this was considered to undermine the exchange programme. In the words of one academic library: "the digitization of departmental series will be the final nail in the coffin for the exchanges". However, opposite views were also expressed. Net publications were considered difficult to use and their future were felt to be uncertain. They could not yet compete with paper publications, which are usable for a very long time. One academic library had struck a nice comprise by making it a part of their service policy to offer both paper and digital publications to customers.

The libraries were asked what they think will happen to the international exchange of publications in the future. Will the international exchange of publications as an acquisition method die in Finland? Only three libraries thought so. A clear majority, 28 libraries, presumed that the activity would continue although some libraries anticipated a further decline. Here too a small difference could be noted between the special libraries whose answers were somewhat more optimistic in tone, expressing hopes that the activity would continue and the academic libraries who stressed the importance of keeping the situation under observation. In both groups of libraries new forms of exchange activity were anticipated and the need for a new way of defining the international exchange of publications was recognized.

Final Remarks

According to the UNESCO handbook, the purpose of the international exchange of publications is to promote the free flow of ideas and scientific information among institutes belonging to different nations and this purpose is achieved by "giving one another publications in printed form or reproduced in some other way" (Handbook 1978, 165). Clearly, in the future, the emphasis will be on 'reproduced in some other way' and 'giving' will imply provide access. As the modes of scholarly communication changes and develops, all initiatives to complement commercial publishing enterprises are attributable to the spirit of the international exchange of publications. The period of transition from paper to electronic formats of publication does not proceed at the same speed in all parts of the world. Libraries, while keeping the situation under observation, will be faced with the challenge of providing different, seamlessly integrated options to their users for a very long time.

References

Barker, Joseph W. (1986) A case for exchange : the experience of the University of California, Berkely. *Serials review*, Spring.

Brophy, Peter (2001) *The library in the twenty-first century : new services for the information age*. London : Library Association Publishing.

Buckland, Michael (1991) *Information and information systems*. New York : Praeger.

Cornish, Graham P. (1990) *Interlending and document supply in Europe*. Paris : UNESCO. 150 s. (PGI-90/WS/13). Also published 1991 as: A European-wide survey of interlending and document supply. *Interlending & Document Supply* 19 (2).

Deal, Carl (1989) The administration of international exchanges in academic libraries : a survey. *Libraries acquisitions : practice & theory* 13 (3).

Dorner, Daniel G. (2000) The blurring of boundaries : digital information and its impact on collection management. In: *Collection management*. Ed. G.E. Gorman. London : Library Association Publishing. (International yearbook of Library and Information Management ; 2000/2001).

FinElib (2003) FinElib Website. [On-line] http://www.lib.helsinki.fi/finelib/english/. Downloaded 6.5.2003.

Fjällbrant, Nancy (1984) Rationalization of periodicals holdings : a case study at Chalmers University Library. *The journal of academic librarianship* 10 (2).

Gibson, Sarah S. (1982) Scientific societies and exchange : a facet of the history of scientific communication. *The journal of library history, philosophy and comparative librarianship* 17(2).

Handbook on the international exchange of publications. Ed. by Frans Vanwijngaerden. Paris : UNESCO.

Herslow Fex, Anna (1983) De commercio litterarum eller hur man kan spara pengar genom att anstränga sig mindre. *Bibliotekariesamfundet meddelar* (1).

Hormia-Poutanen, Kristiina (2001) The National Electronic Library Programme FinElib - the backbone of the information supply in the electronic research environment : universities, polytechnics, research institutes and public libraries forming consortia. *Helsinki university library bulletin.*

Häkli, Esko (1996)Tutkimuksen ja korkeakouluopetuksen tiedonsaannin turvaaminen : kokoelmat ja kokoelmapolitiikka Suomen korkeakoulukirjastoissa. [On-line] http://www.tsv.fi/ttapaht/197/hakli.html. Downloaded 6.5.2003.

Häkli, Esko & Lehtinen, Henna (2000) Doctoral dissertations in the Internet : a good start in Finland. *Liber quarterly* 10.

Järvenpää, Merja (1998) Less money, more management. *Helsinki university library bulletin.*

Kovacic, Mark (1980) Gifts and exchanges in U.S. academic libraries. *Library resources & technical services* 24(2).

Nisonger, Thomas E. (1997) The Internet and collection management in academic libraries : opportunities and challenges. In: *Collection management for the 21st century : a handbook for librarians*. Ed. by G.E. Gorman and Ruth H. Miller. Westport, Connecticut : Greenwood Press.

Sonkkila, Tuija (2000) University libraries and electronic publishing today. *Signum* 4.

Södergård, Pia (1992) *Bytesverksamhetens betydelse för de vetenskapliga biblioteken i Finland : en förstudie*. Åbo. (Meddelanden från Ekonomisk-Statsvetenskapliga fakulteten vid Åbo Akademi. Ser. A ; 362).

Södergård, Pia (1993) The international exchange of publications as an acquisitions method in Finnish academic and special libraries. *IATUL Proceedings* Vol 2 (new series). Also published as: Tarptautiniai Mainai- komplektavimo metodas Suomijos akademinese ir specializuotose bibliotekose. Transl. by Vida Maceviciené. *Tarp knygu* (Vilnius) 412 (1993) 5.

Södergård, Pia (1997) *Litteratur- och informationsförsörjningens nationella nivå : identifiering av orsakerna till Finlands beroende av utländska resurser*. Tampere : Åbo : Oulu, 1997. (Finnish information studies ; 6). Also published as: "Why does Finland make so many international ILDS requests?" *Interlending and Document Supply* 24(1996):3.

Tieteellisten kirjastojen yhteistilasto 1987 (1988) Helsinki : TINFO.

Yu, Priscilla C. (1981) International gift & exchange : the Asian experience. The journal of academic librarianship *6 (6).*

PRESENT STATE AND FUTURE OF THE INTERNATIONAL EXCHANGE OF PUBLICATIONS IN THE NATIONAL DIET LIBRARY IN JAPAN

Hisanori Tanaka
National Diet Library
Japan

Abstracts

As the "national exchange service" of Japan, the National Diet Library (NDL for short) has been conducting exchange of publications, mainly official publications, with most of the countries and regions in the world. With the change of the times, however, we are obliged to review the framework of our exchange service. The international exchange of publications is not only a means of collecting library materials but also a library cooperation activity. And, even in the digital age, its importance still remains as one of the various ways to distribute information internationally. What is necessary in the coming years is to clarify the range covered by the international exchange in our acquisition activities and to develop a more effective acquisition strategy.

Introduction

International exchange of publications is a means of collecting library materials, especially official publications that are difficult to acquire through regular commercial channels. It also has an aspect of an international library cooperation activity, in a broad sense, which promotes "the free exchange of ideas and knowledge among the peoples of the world" as written in the UNESCO Convention. When we think of the future of the international exchange in the digital age, it is imperative that we should return to these two roles and examine them further.

In this report, I will first introduce the current status of international exchange in the NDL on the basis of its history, then raise several points of contention about the present issues and what we should do in the future.

Current Status of International Exchange in the NDL

The NDL has exchange relations with most countries in the world, currently conducting exchanges with 895 institutions of 154 countries and 2 regions and 54 international organizations.

In retrospect, the NDL has been functioning as an international exchange center of publications since its establishment in 1948 when it took over the Japanese government's international exchange project that the Ministry of Foreign Affairs had conducted since 1875. The Library was officially designated for the "national exchange service" when Japan ratified two conventions, namely, *Convention Concerning the International Exchange of Publications* and *Convention Concerning the Exchange of Official Publications and Government Documents between States,* in 1984.

In the legal framework, *the National Diet Library Law* provides that 30 copies of each official publication must be deposited, which enables the Library to keep two or three copies for its own collection and use others for international exchange purposes.

The international exchange of official publications falls into three categories, namely, "blanket exchange", "selective exchange" and "specified exchange". "Blanket exchange" is a system whereby major official publications are comprehensively exchanged based on agreements between governments or conducting institutions. Our present partners are the following six institutions:

1. Library of Congress (USA)
2. National Library of Australia
3. Berlin State Library - Prussian Cultural Heritage (Germany)
4. National Library of Canada (shipped to University of British Columbia Library)
5. University of California at Berkeley (USA)
6. National Central Library (Taiwan)

"Specified exchange" means to exchange basic official publications mutually specified according to an agreement between the governments or the institutions concerned. This exchange is more limited in scale than the blanket exchange. At present, this form of exchange is operated with these 5 institutions: the United Nations Library (in Geneva), the British Library, the Bibliothèque nationale de France, the Russian State Library, and the Royal Library of Belgium.

For exchange with many other institutions, the method called "selective exchange" is taken, that is to exchange materials selected by each other with consent.

Besides official publications, we purchase commercial publications wanted by our partners and exchange them based on the principle of exchange in equal amount or equal price. In particular, exchange of national literature is important. That is, we acquire Japonica, materials on Japan published in other countries, while to those countries we send materials on them published in Japan. This arrangement started in fiscal 1979 with the International Exchange Bureau of Denmark. Since then, the following addressees have been added: the Royal Library of Sweden, the National Library of Poland, the National Library of Romania, the Hungarian National Széchényi Library, the National Library of the Czech Republic, the National Library of Serbia, the National Library of Canada, the National Library of Norway, and the National Library of Korea.

In addition, our Library is designated as the depository library by several international organizations and receives their publications. Since the first designation by the United Nations in 1949, the number of the organizations has increased and decreased. We are now designated by 17 organizations.

The statistical outline is as follows. To begin with, in the movement in the number of exchange partners and received materials, the former has been increasing constantly. On the other hand, the latter reached the peak in the 1980s, after that, showed no fluctuations, and has been decreasing little by little in recent years (see Figure 1).

Then, as regards the details of partners, many of them are in Europe, Asia and North America. Classified by kind of institution, the rate of academic institutions such as universities, governmental agencies, including research institutions, learned societies and associations, and special institutions is high. Especially, we have established a close relationship with the national library of each country. If sorted by country, 129 institutions are in the U.S.A., 56 in China, 47 in the Republic of Korea, 43 in Germany, followed by Britain and then India (see Table 1).

Finally, I would like to outline the NDL's acquisition of materials in fiscal 2001. The rate of publications acquired through exchange among our foreign materials is approximately 15 percent for books and 35 percent for serials, about 90 percent of which are official publications. In recent years, digital materials have been increasing, especially CD-ROMs (see Table 2).

Changes in the Situation of Exchange

1. Change of Framework of Exchange: from Blanket Exchange to Selective Exchange

The blanket exchange system, which dates from the 19th century, had the advantage of simplicity, in an age when each country was not familiar with others' circumstances of publishing. However, when we asked if those materials were actually used and the materials needed by the other were specified, the defect was that many materials for exchange did not meet the needs of the recipients. Through changes of the external environment, such as increase of official publications, addition of partners, reduction of circulation with restricted budget (decrease of deposited publications), rises in postage, and transitions of publishers of official publications to commercial companies, the institutions operating the blanket exchange with the NDL have decreased from 12 in 1959 to 6 at present. With regard to serials, we selected the Core Set Serials based on the demand of the LC in 1995, reconsidered contents sent to the partners of the blanket exchange, and virtually switched to the selective exchange.

2. Decline of Importance as a Means of Collecting Materials

As the result of easier acquisition of bibliographic information of major countries, well- organized circulation of commercial publications and increased number of purchasable materials, the importance of international exchange as a means of acquisition of materials has declined. In former communist countries, because of liberalization of the economy, activities of wholesalers and bookshops increased and we do not necessarily acquire materials via libraries. The defects of international exchange, such as unstable acquisition as compared with purchasing, difficulty in continuous and systematic collection of multi-volume publications and being time-consuming, have become noticeable.

3. Regional Unevenness and a Gap between Supply and Demand

To improve our international exchange services, in 1993, we sent out a questionnaire to 319 major exchange partners to collect data on titles we sent them, titles we received from them, whether they want Japanese language materials and their areas of interest. We received responses from 225 institutions. Many of them wanted to receive materials in English. Those who wanted to receive materials in Japanese were only 39 % (98 institutions), of which 52 were in China, Korea and other Asian regions. In Europe and the US, only institutions of oriental and Japanese studies wanted to receive materials in Japanese.

On the other hand, there was a strong demand for materials written in English. Many institutions' requests concentrated on limited materials such as major statistics, white papers and annual reports. In response to the concentrated demand for specific materials, the NDL had been microfilming about 200 titles of government publications and sending them to 14 institutions since 1977. After seeing the result of the questionnaire, we changed that into microfilming of about 40 English titles and started to send them to the expanded number of 42 institutions.

4. Impact of Digitization of Materials

With the rapid increase in provision of electronic information, such as web sites and CD-ROMs, more and more book form publications are ceasing publication. There are many merits in the transition to electronic information, but libraries need facilities to ensure access to electronic information.

Future of International Exchange

Today, the methods of exchanging knowledge and thoughts have transferred from the exchange of paper media to the circulation of information by electronic media. More and more official publications, such as government information and parliamentary documents, are being digitized and becoming accessible on the Internet. We cannot doubt the fact that the foundation of the international exchange of paper media is shaking.

Libraries must adapt themselves to acquisition of electronic publications and positively promote electronic information access. In fiscal 2002, the NDL has been undertaking an experimental project on the preservation and provision of digital information from websites, and has started to examine how to put networked information into the legal deposit system. We think that the role of international exchange cannot avoid being limited when methods for quicker and inexpensive circulation of information are being established.

However, as I mentioned at the beginning, the exchange service also has a positive meaning of promoting international understanding, as well as being a means of collecting materials. In view of the international information divide, there are still many countries that require exchange of materials in paper form. We have to consider this diversity and act in response to it.

It goes without saying that international exchange has the feature of being one of the most effective means of collecting government publications, doctoral theses which are left out of the commercial distribution process, national literatures, reproduced old materials and microforms. From now on, we need to clarify the areas that the international exchange service should cover and develop a more effective collection strategy.

Finally, we would like to raise the issue of the language barrier from the Japanese standpoint. Only insufficient information has been given in response to the deep interest in Japan from abroad, and the language barrier has long been pointed out. Concentration of demands for limited English government publications is structurally inevitable. The NDL intends to promote the microfilming of English materials in high demand positively as ever. Considering Japan's position in the scene of the world's information exchange, the role of the NDL is important. We will continue to make the necessary efforts to fulfill it.

Table 1 The Items of Exchange Partners

Asia	286	32.0%
North America	150	16.8%
South America	80	8.9%
Europe	272	30.4%
Africa	63	7.0%
Oceania	44	4.9%
Total	895	100%

Table2 Acquisition of Foreign Materials in NDL (Fiscal 2001)

Books			
Purchase	Gift	Exchange	Total
34,845	1,823	6,786	43,454
80.2%	4.2%	15.6%	
Serials			
Purchase	Gift	Exchange	Total
103,240	11,837	59,961	175,038
59.0%	6.8%	34.2%	
Non Book Materials			
Purchase	Gift	Exchange	Total
83,586	216	3,254	87,056
96.0%	0.2%	3.8%	
The Sum Total			
Purchase	Gift	Exchange	Total
221,671	13,876	70,001	305,548
72.6%	4.5%	22.9%	

Fig 1 Change of Exchange Partners and Material

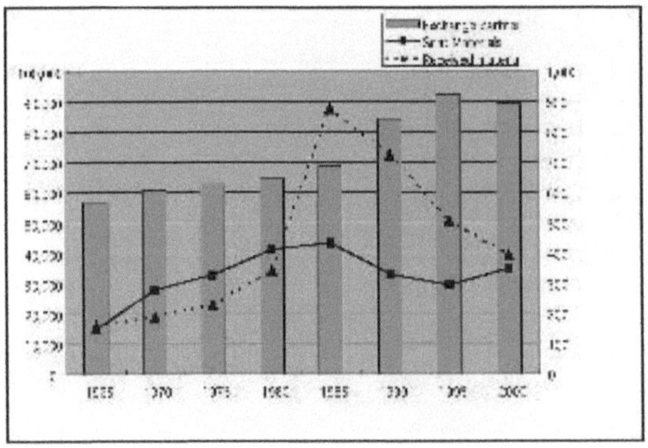

THE EXCHANGE OF ACADEMIC PUBLICATIONS IN SENEGAL

Souleymane Diouf
Central Library, Sheik Anta Diop of Dakar-University
Dakar, Senegal

The exchange of literature as we know it today was mainly established by the 10th General Assembly of UNESCO on the 3rd of December, 1958. UNESCO had accepted two other agreements: the convention on the international exchange of publications, which came into force on the 23rd of November, 1961, and the agreement on the international exchange of official publications, which came into force on the 30th of May, 1961. Our study will deal only with the agreement on the international exchange of publications, and more specific, with academic publications. There has before been an agreement between two institutions, in which each side expresses its will to make known to the partner its publications. This agreement is valid for publications, if the publications either are produced by the institution itself or if the publications are produced by an organisation, for which the institution serves as a repository, without material compensation.

As an academic publication we understand all publications published by the initiative of the university or within its frames (Manuel 1977). Two big categories of documents form the academic publishing on the basis of financing:

- Those, which are financed by the authors themselves, even if that means, that the authors have to find subsidies: these are individual and collective publications like books of authors (scientific in broader sense or pedagogic), proceedings of colloquiums, research diaries, newsletters, conference publications, journals of societies of scientific character and proceedings generally inspired by the universities, thesis and memoirs.

- Those, for which the universities' budgets give financial guarantees: journals with general and specific information about university life and various administrative documents, which don't draw attention to themselves, publications, which could be defined as "official". They are published by various academic institutions and include publications like annuals, newsletters from different academic institutions, which present the work of the teaching staff and researchers; these "official" publications can also include articles received from various sources outside the university or even from abroad, as long as they are of scientific or learned character.

These two categories of documents will be analysed from the point of view of exchanges and the gifts as an act of documentation politics. We use exceptionally the term gift, because if we look further into the future, the exchange with our correspondent institutions will become more and more difficult because of the decrease of our publications and the budgetary cuts, which will have the effect, that donations will play a great role for our holdings. Before a publication can be exchanged or donated, it has first to be produced and evaluated. This is why the first part of our report deals with the difficult situation of production and evaluation of our university's publications in an underdeveloped country as Senegal. After that as a mean of increment and development of the holdings, we will discuss the experiences of the Central Library and the Sheik Anta Diop Fundamental Institute of Black Africa (IFAN) of the Sheik Anta Diop-University at Dakar with some numbers as support. Finally, the second revolution of written documents after Gutenberg's invention of printing, which I would call electronic editing and which has completely changed the process of editing, forces us to reconsider our mission and our methods. The third part deals with the adaptation in exchanges and gifts to the new electronic medium.

1. The Situation and the Evaluating of Academic Publications in Senegal

Among the tasks given to universities, teaching and research are the most important: The task of teaching and training of the human resources the societies need in order to be able to develop; the task to carry out research in order to contribute, advance and develop knowledge. These two activities cannot develop, if their results cannot get published. This means, there is no scientific activity without the actions of evaluating, without editing the research results.

1.1. The Situation of Academic Publications

Publishing in Africa is very difficult. Within the Third World, Africa grants the least amount of human and financial resources to scientific research. It is considered as not immediate profit producing and has therefore no priority. As a result, to the distribution of research results is not paid special attention by our authorities. Before going to the core of this part we would first like to illustrate our report by the reflection of R. Aubrac (1988, 185-191), who has outlined the paradoxical situation of scientific and technical information in these countries in regard to their development. These poor countries do not waste their goods, such as nutrition, clothing, agricultural products or handcraft. On the other hand, their scientific and technical knowledge of their reality and their potential often loses its value in the moment of its acquisition. Despite the fact, that the developed countries take great efforts to produce scientific and technical knowledge and to get this knowledge into circulation, this information, if there is any, is rarely accessible in poor countries. "The rich waste their goods and conserve their knowledge. The poor economise everything, apart of what possibly might help them to escape from poverty " he emphasizes that the knowledge most often is not open to public access: that is, what professor Jacques Mariel Nzouankeu (1993) well notices, when pointing out, that the reason for not publishing scientific works is the lack of means in our universities.

Since 30 years, lots of meetings have been held on scientific and learned publication in Africa. Each time strong and justified recommendations have been made concerning the production, the distribution, libraries and academic libraries. The eternal recurrence of these recommendations is striking. Clearly they have not been translated into action.

As a result of this, the situation of African scientific publication has not improved at all, in contrary, it has decreased. In the Sheikh Anta Diop-University this tendency confirms, except in theses and memoirs, which are increasing in number.

The numbers in the following table have been collected from the yearbook of the university 1972 and from an inquiry at the university.

Years	1940	1973-1992	1993-2000
Periodicals	21 titles until 31st of December 1972	22 titles, of which 9 had an average delay of publication of 5 years. Still 13 regular	10 titles, of which 6 regular and regularly received at University Library. 4 published very irregularly
Monographs (titles available in the holdings)	280	33	32

1.2 The Evaluation of Academic Publications

The libraries have the primary aim to collect, preserve and distribute scientific publications produced by the community they serve and to simultaneously collect the publications in the field of studies of this community. The Central Library is the depository of the publications of the university; this title obliges it to evaluate these publications through conserving and cataloguing.

1.3 Conservation

The task is to arrange the handling of the academic publications as well as for the documentation acquired by the university in such a way, that these publications are preserved in a usable shape as long as possible. The Central Library and IFAN have at their disposal localities and equipment suitable for this purpose.

1.4 Cataloguing

Many academic transactions are being produced, but they often evade from to the National Bibliographic Agency, which the university has to replace in this sector. As a result of this inquiry it came out, that there are great difficulties in identifying academic publications. There is a sort of compartmentalisation, which manifests itself in that, that one faculty does not know, what has been produced in another faculty. Even within a given structure (laboratory, department, faculty or institute), the scientific achievements produced by the teaching staff or researchers are often ignored. A great part of researchers prefer to publish their achievements in the big international journals and in the big publishing houses of the North, which guarantees a minimal delay in publication of the articles. This explains the disaffection towards the local publications, which are much more slowly available. At the moment, apart from the thesis and memoirs, which are deposited systematically, the teaching staff and researchers have no obligation to deposit copies of their publications. There should be an obligation to deposit a certain number of copies, so that it is possible for the Central Library to fulfil its duty of cataloguing and the distribution of the publications on both national and international level.

Despite these difficulties, some initiatives have been made to make the academic publications more visible. The Network of the Documentary Institutions of Upper Education (RIDES) published in 1992 a "Collective catalogue of memoirs, maintained for the University of Dakar and the national schools of upper education, 1957-1981; a collective catalogue of thesis and memoirs available within the documentary institutions of upper education" (with more than 3014 titles). In the sixties, the Central Library regularly reported the situation of academic publications in the form of a booklet and published several catalogues. It is notable that there is no catalogue for the periodic publications edited by the university, because these publications can be counted on the fingers of one hand.

These preliminary remarks on the production, the delay of publication and the absence of rules, which would oblige the teaching staff and researchers to depose some copies of their publications at the library, explain why there is no dynamic exchange policy. Even many of the "official" publications are located in an accidental way. Besides the IFAN, no other organisation has a publication service at its disposal, which would be worth that name.

2. Exchanges and Donations: Acts of Documentation Policy

We have decided to talk at the same time of donations and exchanges for two reasons. Firstly because of the fact, that the mutual donations, appreciated by both sides, can lead as result to a formalisation within this cooperation. This formalisation is then stated by a clear agreement between the two institutions.

Secondly, the shortages of our exchange currency in form of own publications of our university have constrained us from receiving many foreign publications and to accept a very weak counterpart. It has besides never had accountancy to reach a balance in an exchange, which has already been quite out of balance. Donations, which are coming from associations, embassies, foreign universities, private persons... play an important part in the funds of our library. Thus, between 1998 andApril 2001 we received 165 monographs in exchange and 428 as donations. The library limits its activity to direct exchanges, excluding other functions of an exchange centre. In shipment we cannot take profit from any reduction in postal tariffs, but we don't have to pay duty.

2.1 Situation of Exchange

The IFAN has actually an exchange program for both monographs and series principally with the National library of France (Bibliothèque national de France-BnF) and produces continually 3 different series. Between 1992 and 2000, the IFAN produced 19 monographs and 62 different series have been received in exchange for these publications. The library, the audiovisual service and the service of publications form the department for scientific publication. Until 1989 there was a slowing-down in publication activity, which also decreased exchange activity. In 1998, because of this decreasing number of publications, the department set up an exchange policy, which is sufficiently dynamic. The strategy consisted of firstly contacting the researchers to evaluate their needs. After that a list of institutions, which have at least once sent documents was compiled and finally, an old abandoned card index of exchange partners was analysed. This card index included more than 300 correspondents, with which the IFAN exchanged regularly literature. The strategy aims also to supervise publications and to create publication tools in the form of a web site, on which the documents available on exchange are catalogued.

The Central Library, which in the end of 1970 had more than 250 exchange partners, finds itself nowadays in a situation, where it has only 50 exchange partners. The library does not send any monographs, but receives exclusively monographs from the BnF. In exchange, the library sends one of the series for which it acts as depository library. It happens quite often, that we receive a list of publications from the BnF in which we are offered publications for exchange, from which we choose by marking our wishes on the list or by compiling a new list with our choices, mentioning their number on the original list. In the end of the correspondence the BnF often specifies the fields of expertise and the level, in which the BnF wishes to receive publications from us. We are unable to react to these offers, because we don't have any monographs to offer, but we send one of our series, which is available. The reason for this is that there is no obligation for any department, teacher or researcher of the university to deposit copies of their publications systematically.

We exercised earlier the exchange of thesis, but it is not carried out anymore due to lack of financial resources. Medical theses e.g. were initially deposited in 115 copies and were used for broad distribution on international level, but this service ceased to exist due to the lack of money. The number of copies of each thesis deposited in the Central Library has dropped from initial 115 copies to 65, then to 30 copies, 15 and finally to 8 copies.

Today the library exchanges continuously only periodical publications .We have to say also, that the number of these publications has become ridiculous, since, as shown in the table above, the number decreased from 21 constantly until 31st of December, 1973. Only 13 were published regularly in 1992 and finally the number dropped to 6 titles at the moment. These numbers exclude other publications that have been deposited in a systematic way at the library. Only three of these 6 titles serve at the moment as exchange currency for monographs and periodicals.

2.2 Organisation and Methods

A common register for monographs is being launched at the moment for the purposes of exchange and donations. The register includes the following annotations: a register number, the date of registering, the author, the title, the place and date of the work's edition, type of binding (stitched or bound) and a space for comments. This last field is designed for information, if a book has been received on exchange or as a gift. Every partner library has its own file, in which all correspondence with this exchange partner is conserved. The titles of our periodical publications we send in exchange are mentioned on a file card and collected in the file; every copy sent is also noted. There exists a similar file for the periodical publications we receive. The correspondence can have the form of a simple letter, a letter, which includes a detachable reception announcement, a simple shipment declaration form, a sheet of several shutters. It is perhaps necessary to explain here the effective use of printed forms or standardised card indexes because of the international character of an activity like this. In the way of an international exchange format of bibliographic data, a standardisation of set of means must allow a bigger deal of rationalisation, of comprehensibility and of enhancement of operations due to standardised language use. For the moment we are forced to recognise that each partner has its own methods, which are as numerous as there are exchange partners.

This picture we have just drawn, which is not very positive, bases on the on the persistent shortage of publications of our university and on its logic effect on the exchanges for both sides, on the context on general poverty of libraries, whether in north or in south. This empowers the professionals, like us, to clarify some points in order to reverse this tendency.

2.3 Alternatives for the Production and a Better Exchange Policy

The conclusion is that the donations have to be replaced by a real exchange policy. The danger of donations is to receive literature without any value for the university library. The requirements of the development policy of the holdings, in their role of planning and selecting, are contradictory to the compulsions and chances of donations. In the context of a crisis, marked by general diminution in budget, the increase of the number of users and their increasing specialisation, the increasing amount of available documents and in the continuing growth, how should we optimise the exchanges, to make them an efficient mean of entry, which take into account the needs of our users? From our point of view, there is no other possibility than to re-launch the academic publications and to put up regulations with deterrent and binding rules, which makes it obligatory to deposit systematically a certain number of copies of every publication of every single department of the university.

The academic community in its entity must be aware, that our university will by default be removed from the list of publishing centres of excellence and take a not very honourable place, if it is true that the quality of the research activities of the university is measured only by the number publications printed.

I will repeat: the academic publications of a university build the most prominent indicator, the barometer of the productivity and of the spread of a university's research. We will not start again to list encounters on this subject, which is already long enough, but will settle with giving some traces, that can mark the beginning of a solution of the problem, provided these traces will be explored.

- The official publications (journals, revues...) of institutions of the university play an essential role for the career of teaching staff and researchers and for the reputation of the university. Also the specific advantages for those, who accept to publish in the local journals, can be mentioned.

- Many of the proceedings of meetings lay in drawers waiting for the financing of their publication: they could be published in the local revues in form of thematic volumes.

- Giving more money to the University Press of Dakar

- Providing subventions to teaching staff and researchers who wish to publish outside the university and demand as condition a certain number of copies to be deposited at the university.

- The sector of theses has to be explored because it is the only growing sector. The Minister of National education grants subsidies to students writing on a thesis. The university must on its own behalf support these students with additional subventions to graduating students who already are finishing their thesis. This way the number of printed copies of theses could be increased three times higher than now. By sponsoring certain theses, which are judged as excellent in highly qualified fields, it will be possible to wake the interest of the international academic and learned community as well as organs of publication. This is one of the examples that have to be remembered by the university.

The entity of encountered problems and the described usages indicate to which degree the work in exchange is still done manually. In the times of new technologies and networks, what could be the advantages of these technologies for the exchange of publications?

3. The Exchange of Academic Publications and the New Technologies: Which Perspectives?

This last part aims to cause a discussion between professionals representing institutions with different level of technical equipment and therefore is not based on practical experiences. The introduction of the Internet in libraries has had as the most striking effect – the same as everywhere else - the decrease of the delay in communication and the decline of costs; seen under this aspect, the Internet is a mean of communication. Consequently libraries, along with the development of services like GOPHER, WEB and others could offer much better accessibility to the available information. This information may be bibliographic or in full text version. The Internet is also a gigantic reservoir with information easy to exchange.

3.1. Internet as a Means for Communication between Libraries and the Exchange of Publications

Like in many other application fields, the libraries have in the first hand used the Internet as a way of communication. Within the activities in cooperation with other institutions, this is a way to increase the exchanges and to make the communication between libraries more efficient and faster.

- Electronic mail, principally the same as a simple handwritten letter, forms the first application for making the exchange easier between professionals involved into exchange projects. Exchange offers and the search for new contacts can be made much more efficient and much cheaper by using this channel.

- The second step in the utilisation of the Internet as a mean of communication might be professional mailing lists of general interest such as PACS-L or BIBLIO-FR or special interest collected by professionals who share the same interests. Would the time have come to propose the creation of a special list dedicated to exchanges?

- The third step could be the creation of HTML-pages in the frame of libraries' web service, which lists the publications available on exchange, which may be printed, online or on CD-Rom, if necessary, also with a summary on the pages.

3.2 Electronic Publishing and the Exchange of Publications

We may consider all documents, which are digitalised and stored on data carriers as electronic documents. It can be a simple data file, created by text editing and stored on a hard disk of a server connected to the Internet. It can also be pages in HTML-code, paper documents scanned and stored on a data carrier or on CD-Rom. These new document types are integrated into the holdings of libraries along with the traditional holdings. The electronic documents cause often discussions among library professionals on the consistency, the composition and future of the holdings. These documents may be freely accessible or you might have to take further steps to get access to the documents. Such steps might be for example negotiations with a commercial editor, leading to an agreement on a licence between the editor and the user for an access or by passing with a simple password, received by registering to the scientific editor. There is no problem when it comes to the freely accessible documents, because their use is not object to any conditions or restrictions. But how about documents with protected access? In other words, how can we design the exchange of e.g. electronic online-documents in the same way as the exchange of printed documents? In the best possible case this exchange would happen in the World Wide Web for fast access and continuity in publication. Let us use the image of a trade between a commercial editor and the acquisition department of a library, which allows the latter to use an online journal under certain conditions for e.g. the duration of use or the number of downloadable files. The access happens for example by using a password. The same model is normally also usable for libraries which exchange the publications of their universities with each other. This may imply that the universities should be technically equipped with the same publication capacity as commercial publishers. We have a strong belief, that here lies the solution for the crisis of university libraries caused by continual costs of a scientific journal funded by universities and sold to universities by commercial publishers.

Another possibility could be the supply with documents, but this fits not with the idea of exchange of publications, because this would mean the satisfaction on individual demands. That could happen by transmitting parts or the whole document via fax or by sending data carriers. There are several projects in this field: ARIEL, EEDIL, FASTDOC, EURILIA, REDD... This is very efficient work, which shortens the insufficiency in documentary resources in libraries and makes it easier to access these possessions.

The decline of the "order of books" has been proclaimed, but even with the advances in electronics, the exchange of printed publications holds as a possible solution. This is because the licences received from commercial publishers don't allow exchanges. The access is restricted to those persons who work in the institution that signed the contract. Certain documents like governmental publications and grey literature may escape from the commercial circles.

References

Aubrac, R. (1988) Information scientifique et technique (IST) Pamorama general et perspectives historiques. In: Politique, programmation, gestion de la recherché pour le development. Paris, Institut International d'Administration publique.

Fleury, Jean (1973) L'information au service des décideurs. In: CRDI Explore, Vol. 8, N° 3.

Manuel des échanges internationaux de publication (1977). Préparée sous la direction de Frans Vanwijngaerden. Paris, UNESCO.

Nzouankeu, Jacques Mariel (1993) Introduction. In: Le role de l'université dans la production, la conservation et la diffusion des publications universitaires : séminaire tenu à Dakar le mardi 16 mars 1993.

INTERNATIONAL BOOK EXCHANGE: HAS IT ANY FUTURE IN THE ELECTRONIC AGE? A VIEW FROM THE RUSSIAN NATIONAL LIBRARY, ST. PETERSBURG

A.P. Romanov and T.V. Petrusenko

The Russian National Library, St. Petersburg

Russia.

Some History

The book exchange began in the Russian National Library in the late 1950-ies in the context of liberalization of the political regime undertaken by Nikita Khrushchev. Before that time, the single library granted such a privilege was The Lenin State Library of Moscow, currently The Russian State Library, which shares today the status of a national library only with The Russian National Library of St. Petersburg. The rapid growth of the exchange reached its peak in the '60-70s. The exchange became the most important source and at times became even the single avenue for acquisition of foreign books, especially in the field of humanities, as the state funds were designated to subscription of scientific periodicals (mostly medicine and engineering). During the '80s, the volume of exchange began to slide down, reaching its bottom in the early '90s. The significance of this source of foreign book acquisition has however been very high and is growing again. Between 1991 and 1994, it was the only way to acquire foreign books and periodicals.

Although the financing has lately improved, the significance of the exchange is invaluable. It gives us, in the recent years, around 30 000 items of periodicals, 7 000 books, let alone newspapers, posters, and so forth, coming from 1 150 partners in 77 countries.

Positive Aspects

The international book exchange viewed from The Russian National Library (and perhaps from the broader Russian perspective) looks like an affordable and effective source of acquisition. Current Russian budgetary difficulties and the small portion of the budget allotted to cultural institutions augment the benefits of exchange in comparison with purchase. Even discounts that book traders give to libraries do not bridge the price gap: Russian books, which we exchange, are a lot less expensive, that is, through exchange, we are buying foreign books by prices, comparable with Russian ones. Even the growing postal tariffs in Russia, which are approaching international ones, still have not erased the value of book exchange.

As a result of sociopolitical changes, the present Russian book market is richer and more diversified than ever before. The Russian reader has never had access to such a variety of books, many of them belonging to the category proscribed before the end of 1980s. These are in the first place books of social sciences, history, belles-lettres and so forth. The editions are numerous but small, printed by little, often short-lived companies and are difficult to obtain except by book exchange.

Internal Threats

The most obvious difficulty we face today is the low budgeting of Russian libraries. While we don't know how long this trend will last, we are well aware of the imminent threat of ever growing prices of periodicals and books in Russia. So far, the prices are not nearly as high as the average ones on the global scale. Their growth is held down by the low purchasing power of population. Yet the tendency is too clear to be ignored.

An exacerbating factor is the state of tax laws that threaten the foreign book acquisition. According to the tax bill, which was long under discussion in the Duma, Russian libraries pay added-value taxes on foreign editions beginning with 2001. Finally, after changes and reductions, in 2003 the VAT represents 10% for books and periodicals. The book exchange material now is free of VAT, but this problem has different aspects and it should be considered in a special paper.

An additional burden is imposed on libraries by time-consuming customs procedures. Despite of some positive changes in this matter, these may often be, if not very expensive, but too labor-consuming. Smaller libraries simply cannot handle them and have to pay top money to agents or are forced to abandon book exchange altogether.

The most troubling problem that looms large on the horizon is associated with the problem of human resources, which are quite different in Russian libraries than in the Western libraries. The first problem in this regard is the growing shortage of qualified specialists. From the outset of the business of book exchange, the service engaged highly qualified specialists, mainly Leningrad university graduates in philology and history. They had command of several foreign languages, they were intimately familiar with fine literature, were experienced in working with bibliographical sources, and were very often themselves the first and very much involved readers of the books they acquired for the library. In the early '90s the situation drastically began to change. The sociopolitical changes Russia underwent effected dramatic change of values and professional orientation among young people. These days young university graduates have opportunities they didn't have in the '50s and '70s. A lot more lucrative jobs than those in libraries have opened for them. At the same time, the salaries of librarians have dropped so low that they are today awfully inadequate. Educated people cannot any longer afford themselves interesting jobs.

Today, The Russian National Library employs 20 librarians in the division of foreign book exchange. Such a team seems to be a luxury by the Western standards, but in fact we cannot any longer sustain the high quality of the staff. Many of those who are well qualified for the job with foreign exchange and acquisition, have already reached the retirement age, and there is no replacement for them from the pool of sufficiently educated people. This is a threatening situation with no solution in view.

Outer Threats

Manifestations of the outer threats have been there for quite some time. The first one was noticed already in the 1980s: dwindling down of the interest for the Russian book has been registered in many countries. Our main partners in the book exchange are today mainly large university libraries in the West which keep maintaining of Slavic collections high on their list of priorities. The number of such libraries is limited, and their orders have significantly diminished in comparison with the '60s-'80s.

If in the '60s, there were practically no other sources for the purchase of Soviet books than The Interbook (Mezhkniga), a monopolistic state agency, today there is a rather large net of competing book-selling companies, which have penetrated Western markets. As a good example, I can mention the publishing company Dmitry Bulanin of St. Petersburg which has established well working connections with various libraries and booksellers in Europe and America. The activities of such companies still are spotty and uneven in various countries, but their presence becomes steadier, and their mutual competitiveness brings down prices of Russian books in the West. This tendency is visible quite well in Germany. We have no data, but it would be safe to assume that similar processes are under way in the US.

Clearly, the situation of the Western libraries is different and is ridden with different but nonetheless severe problems. The book exchange business requires knowledge, analytical skills and is labor intensive. It needs a work force with very special qualifications. It requires a team work, and the teams can neither be too small. As far as we know, under the present time budgetary limitations, our partners are insufficiently staffed. And having no capacity to check out carefully the lists of supply against the demands of their respective institutions, they are forced to reduce their book exchange (we suppose, one can argue that this leads to stricter criteria and to a more focused selection). When the books exchange cannot be handled properly, purchase of books becomes the alternative of necessity.

As we have learned from the representatives of Western book-selling companies, the above mentioned tendency compels libraries in the West to devise new strategies. Increasingly often, European and American libraries commission the booksellers not only to buy books for them, but also to do selections in accordance with criteria spelled in cumbersome itemized forms. Such services are expensive, but employing teams of qualified specialists in various fields is even more expensive.

Finally, new possibilities opened by the cyber age are not to be missed. The rapidly growing availability of electronic copies of documents housed in libraries makes the physical accessibility of a book or a periodical unnecessary. Russian libraries have today electronic means at their disposal. Electronic copying concerns in the first place documents needed for narrow circles of scholars and scientists. Electronic copies serve this kind of readership well, especially in the case of papers from periodicals and parts of books.

Even before the arrival of amazing contemporary means of telecommunication, already in the '70s, we registered a decrease of orders for periodicals from our exchange partners, because scientific and scholarly magazines became available through well coordinated channels. This tendency was especially clear in regard of publishers of magazines that conducted exchange with us. If in the '60s through mid '70s their part in the book exchange was significant - we used to have many hundreds of such partners, - today, their participation is very modest.

Means of the Electronic Age

It is hard today to forecast the possibilities which will open in ten years from now. Even in the nearest future, we may expect surprises. What seems to be fantastic today will shortly appear in the field of telecommunication, electronic delivery of documents and technologies of making electronic copies. Currently, procedures of document scanning are time consuming, and their feasibility is limited to documents of relatively small size. There is no doubt that these technologies will become more productive in the near future and thus will take a larger share in the international book exchange.

The International Book Exchange: Traditionally Produced Documents and Their Electronic Copies

Then how can we envisage the future of the international book exchange? Will it withstand the onrush of new technologies or is it bound to become a turned-over page in the history of foreign book acquisition? The overall tendencies are of dual nature: on the one hand, it is clear from the experience of the last decade that the book exchange is dwindling down; on the other hand, it becomes more focused and better fitted to the special needs of partners. We would like to underscore that notwithstanding all the progress in electronic technologies and proliferation of new modes of delivery, libraries still overwhelmingly prefer obtaining traditional paper copies (!) of books and periodicals. We may venture a prediction that the book in its traditional form would acquire in the future even greater value in the eyes of erudite researchers and book lovers in comparison with our own time. A book carries not just information embedded in the text and illustrations, but also a multitude of imprints and testimonies of its age which become more eloquent with the passage of time and usually are not very discenable in its own time. These testimonies will be largely erased by the electronic means of copying no matter how perfect they could become and not carried over in full to the users of future. Yet these testimonies are part of cultural legacy. No doubt, electronically made copies will become a routine component of the book exchange. The proportion of it will stay in direct association with the profile of a library as well as with the availability of funds. The more unique a library is, the more capable it will be to supply its partners with copies of documents and to exchange them for other documents in any form — traditional or digital. Of course, the international community of librarians will have to solve numerous problems concerning the copyright and licensing. The exchange of editions published in small numbers or those distributed outside of the commercial nets will have its adherers under any circumstances.

A Final Remark

We would like to point out one particular feature of organization of the book exchange in The Russian National Library. From the very beginning, this field attracted specialists of high qualification who, besides the routine library work (conducting correspondence with partners, keeping an inventory of exchange, and so on), were commissioned to monitor the publication scene in every particular field of scholarship that corresponded to the field of their initial specialization acquired at university. This mainly concerned literary scholarship and linguistics, while engineering and medicine were in the hands of specialized bibliographers. This way, the international book exchange stimulated the formation of a group of specialists capable of solving various problems associated with compiling collections of foreign books in various fields at the level suitable to a national library. Such conditions have allowed us effectively to conduct the international book exchange as well as other forms of international activities.

LIBRARY EXCHANGES IN OUR CHANGING WORLD: REVIVING THE DISCUSSION WITH SNAPSHOTS OF EXCHANGES IN FINE ARTS, UNIVERSITY AND SPECIAL LIBRARIES IN THE USA AND CANADA

Joyce Pellerano Ludmer

Research Library, Getty Research Institute

USA

During the past decade, as libraries and information services adjusted to the dynamic automated environment, many terminated their exchange programs, citing staffing and fiscal issues. In an effort to encourage reviving the discussion on book and serial exchanges[1], this article aims to explore the existing situation, citing examples of current practice, culled from informal conversations with librarians working in key American art, university and special libraries. But more is asked than answered in this short space. And these major questions remain as suggestions for our continued discussion: Are library exchange programs relevant, useful, perhaps even necessary, in our changing world? Can they be cost effective? What role can IFLA play in exploring the issues?

Exchanges in the 19th and 20th Centuries

Exchanges began about one hundred and seventy-five years ago in American academic libraries and research institutions. Around 1825, American research libraries began exchanging serials and books with librarians in the Russian Empire, and these exchanges continued with few interruptions even after the revolution of 1917. (Lorkovic and Johnson 1997, 59-87). Most of these Russian and Slavic programs still exist today, despite institutional budgetary constraints and the increasing unreliability of several foreign exchange partners. A more general exchange program began about one hundred years ago at the University of California, Berkeley. This program grew into one of the largest exchanges developed in the USA. It continued to expand in size, scope and cost until the 1980s when the program employed five FTE staff, handled over 15,000 active book and serial titles, and managed a materials budget of $150,000. (Carothers 1999, 97-108.) The 1990s brought budget cuts and reductions. And in 1999, Berkeley redesigned the program to one slightly smaller in scope and "substantially more cost-effective than its predecessor" (Ibid). About five years ago, the program continued to

---

[1] For a good current definition of exchanges: Carothers 1998, 32. See also, Lorkovic and Johnson 1997, 61.

bring in difficult-to-locate monographs (approximately 4,500 annually) and over 12,000 serials (Ibid 97). At about the same time, librarians at Berkeley's sister institution, the University of California, Los Angeles, had reached a different decision (McKinley 1986, 75-80). Because the UCLA librarians' review of serials exchanges found their program less cost effective, exchanges were stopped, with the exception of those few serials available only through exchange.

During the second half of the 20th century, many varieties of libraries and information centers offered book and serial exchange programs, each differing in size and scope. Some programs focused on serials alone while others collected monographs, using blanket or barter exchanges, depending on institutional needs and partners' strengths. But whatever the collection or the approach, most institutions found it necessary in the 1990s to reevaluate their exchange programs and consider difficult cost-based decisions.

Exchanges in the 21st Century

This informal review suggests today's existing pattern, revealed through conversations and email communication with a group of colleagues.[2] In the first group are a number of institutions that terminated their exchange programs in the 1990s. Some art librarians now are concerned that these collections may lack specific publications, such as serials and books from East and Central Asia or art exhibition catalogues from small foreign regional museums and art galleries. One institution had ended its program with over 240 exchange partners but is now considering initiating a small focused foreign exchange for monographs and serials.

The second group includes university and special libraries that decided to streamline their exchanges to improve efficiency. These libraries continue to support thriving programs that are heavily labor-intensive but help to stretch acquisitions funds and enrich collections. One institution has about 140 exchange partnerships

---

[2] My warmest thanks to my colleagues at the following institutions who responded to my questions and shared their expertise: Jack Perry Brown, Director, Art Institute of Chicago, Ryerson-Burnham Libraries; Terence Ford, Head, Research Databases, Research Library, Getty Research Institute; Luke Swetland, Head, Information Resources, Getty Conservation Institute; Amanda Bowen, Collection Management Librarian, Fine Arts Library, Harvard University; B. J. Kish Irvine, Ph.D., Indiana University, Fine Arts Library; Daniel Starr, Manager of Bibliographic Operations, Metropolitan Museum of Art, Thomas J. Watson Library; Jo Beglo, Bibliographer/ Bibliographe, National Gallery of Canada / Musée des beaux-arts du Canada, Library and Archives; Alex Ross, Head Librarian, Art and Architecture Library, Stanford University; Frank Carothers, Gift and Exchange Librarian, University of California, Berkeley, University Library; Deborah Griffith Davis, Bibliographer for Music, Acting Bibliographer for Art and Cinema, University of Chicago, Regenstein Library; Susan V. Craig, Head, University of Kansas Art and Architecture Library; Laurel Bliss, Librarian for Art and Architecture, Yale Arts Library.

with foreign museums and libraries; another continues exchanges with 100 libraries. A third library unhesitatingly has pursued its extensive program without a break, but reports that several of its large East European and Asian exchange partners were forced to cancel their partnership between 1989-1995 because they could no longer afford to maintain their plans. Several research libraries with significant Slavic collections considered the idea of stopping their traditional serial and book exchanges to rely only on commercial vendors but, in almost all cases, decided to continue their programs.

In each institution, librarians maintained careful oversight of their programs and immediately cut back on those exchanges where partners had listed prices that were found to be more expensive than commercial vendors' quotes. Most American libraries stopped domestic exchanges. Smaller libraries scaled back and looked to their directors, art librarians, bibliographers or subject specialists to arrange for exchanges in specific clearly-defined fields, usually only with foreign partners. Other libraries stopped their organized exchange programs but still receive numerous materials through campus museums, which continue exhibition catalogue exchanges with sister museums.

Most colleagues underscored two major advantages offered by their exchange programs. They noted the strengthening of research collections by adding difficult-to-find serials, monographs and short-run art exhibition catalogues.And many stressed the immeasurable importance of an exchange program in the presence it gives the institution. Everyone realizes the fiscal issues and continues to search for ways to address the persistent costs needed to staff and run these programs. However, most libraries have collected only in-house statistics, with no comparative data from other library programs, to help them reach informed decisions.

What about the Future?

Various types of libraries and research institutes have specific interest in collecting illusive publications, including so-called "grey literature." Serious art research libraries, for example, search for art exhibition catalogues published all over the world by small regional museums and galleries. These publications, often in small runs, rapidly disappear and are infrequently handled through normal trade channels, particularly publications from areas such as Latin America, East Asia, Central Asia, the Near East and countries from the former Soviet Union. Special libraries also need to collect publications, often hard to find, that describe cultural artifacts, not only our own but those from other cultures, and to document restoration projects of art objects and architectural monuments.

Exchanges assure a library serials that are difficult to find and materials published by organizations, possibly those in developing countries, which need books and serials from other libraries but cannot afford them.[3] One library administrator noted that, in his experience with countries in South Asia and Latin America, subscriptions are often stronger and more dependable when obtained via exchange than by paying subscription costs to vendors. And also, library exchange programs are often the only way to assure consistent receipt of these publications.

The question of cost looms on the horizon for all institutions considering exchanges. Depending on the size and scope of the institution, exchange budgets range from $124,000 down to $10,000. Some smaller libraries receive titles from existing local museum programs. Other libraries absorb costs under the acquisitions budget, realizing titles are added that cannot be found in usual trade channels or noting appreciable savings when compared to titles available through vendors. The librarian of one major art library notes that their exchange program had run at about $12,000 annually, but was cut this year and would face the overall 15% reduction were it a budget line item. Because the librarian and administrator recognize its importance in the development of the collection, the exchange program continues, but its support will come from gifts and fund-raising efforts, until the time it can return as a line item on the institutional budget.

At this point, each library is following its own path. In several research libraries, collection development librarians and subject specialists are establishing individual contacts with exchange partners, foraging a crop of titles desired for their own collections. But the issue of cost effectiveness needs bottom line data for true and valid assessments to bolster informed administrative decisions.

Joining Forces under IFLA

Librarians could consider joining forces, perhaps under the aegis of IFLA's Acquisition and Collection Development Section or IFLA's Special Libraries Division, Section of Art Libraries, to foster the development of an informal committee focused on collecting data in a systematic way. The study might begin by focusing on one specific subject area. The committee might initiate this study by reviewing and updating data already collected by individual institutions. For example, Ohio State University Libraries did a detailed statistical study in 1997, discovering by the end of the project that "overall costs for purchasing titles received on exchange were not significantly higher than the costs of titles that were purchased previously." Most of the titles that they retained on exchange were published by special societies and by university presses in the Far East and Pacific Rim countries and former Soviet Union nations. And they found that foreign exchanges for these difficult-to-find titles had a "decidedly higher reten-

[3] See the Organization for Economic Cooperation and Development (OECD) at www.oecd.org; United Nations Development Programme at www.undp.org; and Association for Information Systems at www.is.cityu.edu.hk/research/resources/isworld/developingcountries/index.

tion rate" especially in areas with volatile currency. (Diedrichs and Davis 1997, 377, [373-378].)

The committee's data collecting could focus on identified geographic areas where published materials are difficult to acquire. Information is needed on the number of series and serial titles available only through exchange. Another vital factor is the number and type of scholarly research material required in special libraries but lost when exchanges are terminated. Lists can describe publications from certain organizations that are more easily available through standing order plans. Another relevant issue involves the proliferation of e-journals and its effect on serial exchanges.

As important as it is difficult, one other question needs our focused attention: how can a library measure the value of worldwide institutional presence against bottom line issues? And tied to the idea of institutional presence is an altruistic impulse. Published in 1997, but still fresh and important today, is this moving decription of "The Philosophy and Appeal of Exchange" (Lorkovic and Johnson 1997, 61-62).

When Frenchman Alexandre Vattemare lobbied the U.S. Congress and various European governments in the 1830s to formalize the international exchange agreements of official publications, he argued that exchanges were a means 'to open the channel of communication between the people of various nations of the world, ...and by making them better acquainted with each other's laws, manner and customs and intellectual wealth ... to cultivate the spirit of peace and of reciprocal respect and good feeling...' (Ibid.)

The 19th century regulation of official exchanges was quickly followed by unofficial or barter exchanges between libraries among a growing number of nations. One must wonder if Vattemare's words and the early altruistic appeal of exchanges can return as a constant to today's deliberations.

Without doubt, the 21st century library must put cost effective issues center stage in the review of exchange programs. But in order to reach cost effectiveness, does a library have to terminate exchanges? Or do research libraries benefit most from a streamlined approach to exchange programs? After gathering data on those materials easily acquired through vendors or institutional standing orders, librarians can identify publications available more readily through exchanges. Small exchange programs focused on geographic areas, such as the Far East, Near East, Central Asia, the former Soviet Union and Latin America, assure that research libraries add serials, monographs, and art catalogues often lost in the normal acquisitions process. Collection development needs and efforts to strengthen a research collection are vital considerations that librarians and subject specialists must definitely include in their cost effective studies. And librarians must find ways to factor into cost discussions the evasive idea of institutional presence. We return to our question: Are library exchange programs relevant, useful, perhaps even necessary, in our changing world? Only with further focused discussion of the current exchange situation we can reach wise, collection-strengthening, yet cost effective programs for our libraries and research institutions.

References

Carothers, Frank (1998) Gifts and Exchanges. In *Technical Services Today and Tomorrow* by Michael Gorman and associates. 2nd edition, edited by Michael Gorman. Englewood, Colorado: Libraries Unlimited, Inc.

Carothers, Frank (1999) Procedural and Programmatic Responses to Budget Cuts in Berkeley's Exchange Program. In *Gifts and Exchanges: Problems, Frustrations, ... and Triumphs*, edited by Catherine Denning. Co-published simultaneously as *The Acquisitions Librarian*, no. 22 1999. New York, London and Oxford: The Haworth Press.

Diedrichs, Carol Pitts and Trisha L. Davis (1997) Serials Exchanges: Streamlining and Elimination. In *The Serials Librarian*, 31:1/2, 1997.

Lorkovic, Tatjana and Eric A. Johnson (1997) Serial and Book Exchanges with the Former Soviet Union. In *The Serials Librarian*, 3 (4), 1997.

McKinley, Margaret (1986) The Exchange Program at U.C.L.A.: 1932-1986. In *Collection Development and Administration* 12:1, Spring 1986.

PART THREE: DIRECTORY

ORGANISATIONS AND FUNCTIONS OF THE EXCHANGE CENTRES WITH A NATIONAL RESPONSIBILITY

Ed. by Sari Lehtinen

Exchange Centres with a National Responsibility

Exchanges can be handled either directly between libraries or through national exchange centres. If a country has no exchange centre, each institution wishing to exchange publications with institutions in other countries has to do this on its own, i.e. it chooses its partners independently, ships and receives letters and parcels directly, and bears all expenses incurred in this work. If, on the contrary, one or more exchange centres have been established in the country, it is up to each individual library to decide whether it will make use of some or all the services of the centre or carry out exchanges directly.

1.National Exchange Centre

The first Brussels multilateral Convention on the International Exchange of Official Documents, Scientific and Literary Publications of 1886 postulated the formation of a "bureau charged with the duty of exchanges" in each of the contracting States, thus officially introducing the notion of the national exchange centre into the exchange practice and creating a legal basis of the practice of its existence. In the following decades many exchange centres were established in the States which concluded, or adhered to, the Brussels Convention as well as in other countries.

The Paris Unesco multilateral exchange conventions of 1958 reiterated the necessity of establishing national exchange services or other central exchange authorities although this demand was far less categorical than the requirements of the first Brussels convention. Between 1960 – 2003 47 states ratified/ accepted/ notified of succession the Convention concerning the International Exchange of Publications.[1]

The Article 3 of the Convention formulates the national exchange service and its main tasks:

The Contracting States may entrust the national exchange service or, where no such national exchange service exists, the central exchange authority or authori-

[1] http://erc.unesco.org/cp/convention.asp?KO=15395&language=E

ties with the following functions in connexion with the development and co-ordination of the exchange of publications among bodies and institutions referred to in Article 1 of the present Convention: (a) facilitating the international exchange of publications, in particular by transmitting, when appropriate, the material to be exchanged; (b) supplying advice and information on exchange possibilities for bodies and institutions at home and abroad; (c) encouraging, when appropriate, the exchange of duplicate material.[2]

The other convention of 1958 is the Convention concerning the Exchange of Official Publications and Government Documents between States. Between 1959 – 2003 51 states ratified/accepted/notified of succession the convention.[3] In the Article 4 there are paragraphs concerning the exchange centres:

1. In each Contracting State, the national exchange service or, where no such service exists, the central authority or authorities designated for the purpose shall carry out the functions of exchange.

2. The exchange authorities shall be responsible within each Contracting State for the implementation of the present Convention and of bilateral agreements as referred to in Article 3, whenever appropriate. Each Contracting State shall give its national exchange service or the central exchange authorities the powers required to obtain the material to be exchanged and sufficient financial means to carry out the functions of exchange.[4]

During the period of 40 years the amount of national exchange centres has grown. The organization and functions of the exchange centres vary considerably from country to country. In most cases the exchange centre is attached to the national library or to other kind of respective library. Some exchange centres serve all libraries and exchanging bodies of the country. Majority of the centres serve only a part of the libraries (e.g. the national library, university libraries with their branches or the libraries belonging to national academies). During the last years some exchange centres of national responsibility have discontinued their activities (e.g. Denmark and the United Kingdom).

2. Organisation of the Exchange Centre

The exchange centre with a national responsibility may be either an independent body controlled only by a ministry or by some other governmental agency or it

---

[2] United Nations Educational, Scientific and Cultural Organization: Records of the General Conference, Tenth Session, Paris 1958. Unesco 1959, pp. 87-88 (also http://unesdoc.unesco.org/images/0011/001145/114584e.pdf)

[3] http://erc.unesco.org/cp/convention.asp?KO=13036&language=E

[4] United Nations Educational, Scientific and Cultural Organization: Records of the General Conference, Tenth Session, Paris 1958. Unesco 1959, p. 90 (also http://unesdoc.unesco.org/images/0011/001145/114584e.pdf)

can be attached to another institution (e.g. national library). In this case, it can either be only a section or have some varying degree of autonomy. No strict rule can be laid down in this respect because the varying conditions in the individual countries make this impossible. However, an exchange centre is in fact a comparatively small unit so that it is preferable not to make it too independent but give it the backing of a larger institution, which should not, of course, entirely absorb it. Thus, some sort of financial autonomy could be considered as a satisfactory solution.

The exchange of official publications is conducted between States, not between individual libraries. National exchange centres dealing with official publications, whether or not they are empowered to conclude individual agreements, are always guided by the government's general instructions. The choice of exchange partners is often made by the respective government agencies, but even if it is entrusted to the centre, the centre is obliged to act strictly in accordance with the established government policy. The exchange of official publications is usually entrusted to one organization in a country. It can be the national exchange centre which is in charge of other exchange publications or it can be another organization (e.g. the parliamentary library). A profound review to the concept of the international exchange of official publications and carrying out by the German Exchange Centre (Berlin State Library) was presented by Johannes Metz at the 65th IFLA Council and General Conference 1999. [5]

3. Functions of the Exchange Centre

3.1 Provision of Information on Exchange Possibilities and Facilitating Exchanging Bodies

The first and most significant task of an exchange centre with a national responsibility is to provide information on the exchange possibilities of domestic and foreign institutions.

The centres may give information of the material available in exchange (e.g. lists in Internet or paper lists sent to both domestic and foreign exchange partners). A database including the addresses of active and potential exchange partners and the material sent and received in exchange and the material available on exchange is very important and helpful to the exchange centre and the cooperating libraries. It is also recommended to provide a web portal for helpful directories and catalogues.

---

[5] Metz, Johannes: International Exchange of Official Publications. In: *INSPEL 34(2002)2. Washington 2002, p. 80 - 89* (also http://www.fh-potsdam.de/~IFLA/INSPEL/00-2mejo.pdf). Paper presented at the 65th IFLA Council and General Conference Bangkok, Thailand, August 20 – August 28, 1999. Section: Government Information and Official Publications

The centre may also assist libraries in one's own country in carrying out direct exchanges or it can be in charge of the whole exchange process of libraries or other institutions by making direct exchanges on behalf of other libraries (taking care of the correspondence and agreements, offering actively publications to exchange partners, delivering and controlling the exchange material).

Exchange centres are also handling direct exchanges with foreign exchange centres or individual libraries by having sufficient exchange material (legal deposit, collection of duplicates, regular gifts, book-trade, copies of research publications granted by the issuing bodies, etc.) at their disposal, acquired through various channels. In order to avoid double work, the exchange centre should cooperate with other libraries maintaining direct exchanges. Thanks to cooperation between libraries and new library systems with similar or shared electronic catalogues the libraries can work together and also coordinate library acquisitions. It is appropriate to coordinate exchanges as a part of acquisitions in some level for instance 1) to help the foreign partners to acquire required material from the different available reserves in a country and vice versa, 2) to look after that publications in all fields of scientific research and geographical areas can be received, 3) to control that the exchange material is in a suitable collection, 4) to avoid too many duplicates of foreign publications received in exchange (if not necessary).

However the coordination must not limit the freedom of an individual library in practicing its exchanges.

3.2. Forwarding of Exchange Material

A very important task for the exchange centres with a national responsibility has been the centralized transmission of the exchange material of individual institutions. By forwarding the exchange material, the exchange centre saves the individual institutions costs and labour. It spares the trouble of studying duty and export regulations. The Unesco conventions give exchange centres a better chance to exemption from customs duties, to have a discount of postage and taxes than the multitude of libraries in a country.

However collective consignments are often too slow in delivering the exchange material. Nowadays many exchange partners prefer exchange material that is received relatively rapidly after their publication (especially concerning periodicals). A bulk of publications in one shipment means the receiving partners in many countries also a lot of work and duties with the customs authorities. Majority of the centres don't send exchange publications in big parcels or by slow transportation (e.g. sea transport) any more even if they are in charge of exchange transportation of more than one library. It also seems to be the trend that exchange centres don't send exchange publications in such large quantities as before, but try to serve a more limited clientele instead. It is also customary that libraries in their role as exchange centres use the materials they have received or acquired for their own exchanges. A similar change has taken place in regard of the receipt of foreign exchange materials: many exchange centres accept only materials, which are sent to their own library. The economic efficiency of joint consignments was questioned already in the 1970's.[6] Joint consignments require large provisional storing facilities, unpacking and repacking and careful documentation. There are however, various solutions regarding the form of activity, depending on the other tasks of the exchange centre and on its organisational structure.

3.3. The Intermediary Role in the International Exchange of Duplicates and Surplus Material

This intermediary function of national exchange centres has been assumed by some centres since 1955 when Unesco (which had been carrying out this work on an international scale during the first decade of its existence) transferred it to them. Its original aim was to help to fill the gaps in library collections left by the war. Now it is an important means of completing sets of periodicals or series, acquiring old and out-of-print titles, etc. This intermediary function stimulates the exchange of duplicates within a country, thus easing the burden of surplus material, and makes the duplicates available to individual libraries or to exchange centres for international exchange purposes. The role of the exchange centres may be limited to the distribution of lists of offers or requests from domestic and foreign libraries. On the other hand libraries can send duplicates to the exchange centre directly. The exchange centre stores them up and maintains and sends exchange lists either in paper form or provides them through the web.

The lists of offers should be classified on the basis of the subject, so that they may be sent only to those interested, or if the lists are in the web, to make the search easier. Lists can be sent either to one library at a time or simultaneously to all libraries at the same time. Libraries may choose the items they need and return the offers along with their requests within a given period of time. If the list of duplicates is in the web, the duplicates should be available as long as they are kept visible. There is usually only one available copy of each item listed and it is up to the centre to decide to which exchange partner this copy will be forwarded.

It is useful to keep a record, what material is being sent, what the wishes and needs of the libraries concerning materials they wish to receive are and what materials has been sent.

---

[6] I. Gompocz, 'Economic Aspects of the International Exchange of Publications', op. cot., p. 267 – 81; P. Genzel, 'The Efficiency to Collective Consignments', in: Marial Schiltman (ed.), *The International Exchange of Publications*, op. cit., p. 66 – 73; P. Genzel, *The Efficiency of Collective Consignment, Final Report*, IFLA Committee on the International Exchange of Publications, fortieth session, Washington, D.C., November 1974

The exchange centre can store duplicates or other surplus materials it has received from abroad and send them to libraries upon their wish. Certain centres have developed into duplicate exchange centres which centralize the exchange of duplicates for the whole country (e.g. United States Book Exchange (USBE)[7]. Such a centre needs not be identical with the national exchange centre; it might be an autonomous section of it or even an independent body.

3.4. The Future Function of the National Exchange Centres

Even so the importance of the national exchange centres seems to be smaller than before and in some countries there is no reason not to establish or to develop the activity of an exchange centre if it is regarded beneficial. The significance of the exchange centres as intermediary dispatching organisations has diminished Nowadays exchange centres have direct exchanges and keep collections of materials designed for exchange. Depending on the country in question the centres keep all the materials they receive in their own library or distribute them to various libraries in the country. The partial transfer to electronic form of publishing will also have an impact to the role of the whole exchange activity in the future. Thanks to the electronic form it will be easier to distribute and to access information, but purely commercial publications will remain outside of the scope of the exchange activity.

There are some positive aspects if a specialized organization is in charge of the international exchange: 1) the maintaining of special exchange databases with web services is worth and makes the knowledge needed in exchange possibilities and operations accessible for all libraries in a country, 2) it is important to be aware of and to receive the material that can only be obtained within the framework of the international exchange of publications, 3) the exemption of duties and discounts of postage and taxes are possible and thus save expenses, 3) the routines of the exchange can be performed with a small amount of labour force, 4) it is possible to coordinate the exchanges, material received in exchange and to count the economic profit. The most advantageous situation is when the exchange centre is a part of a library or of some other similar organisation. The flexible organization and a close cooperation with libraries and publishing bodies help the work.

All domestic, as well as foreign exchange centres and libraries practicing exchange and other kinds of organisations, should be kept informed if the activity of exchange centres changes or if new centres are being established. Because the organisation of the Unesco has changed, Unesco is no longer keeping record of national exchange centres and their activity.

---

[7] http://www.usbe.com/

LIST OF EXCHANGE CENTRES AND LIBRARIES WITH NATIONAL RESPONSIBILITY*

ALBANIA

National Library of Albania
Palce Scanderbeg
Tirana

Tel (355 42) 23 843
Fax (355 42) 23 843

The National Library of Albania has book lending and exchange relations with various libraries in Europe and elsewhere.

The National Library of Albania keeps and develops relations with other counterpart libraries and institutions, and works for memberships and participation in the international activities. Following the democratic changes in Albania in 1990, the National Library of Albania has become a member of IFLA, LIBER, CENL, CDNL, ELAG, GABRIEL, etc. Based on the bilateral and multilateral agreements as well as on international memberships and the traditional relations with other institutions and libraries, the National Library of Albania strives to have a better cooperation and exchanges with other countries. The direct contacts are greatly estimated, because they open up and create new possibilities for all Parties.

* The information is collected by the Finnish Exchange Centre for Scientific Literature. 66 countries did not answer to questionnaires and 12 of the countries included in this edition did not exist when the 4th edition of *Handbook* came out. The descriptions are published in the form written by each institution.

ARMENIA

National Library of Armenia
72, Terian Street,
375009 Yerevan

Tel (374 1) 58 42 59
Fax (374 1) 52 97 11
nla@arm.r.am
http://www.nla.am

The Book Exchange Department of the National Library of Armenia has a staff of five full-time employees. The Department consists of two sections: local (inland) and foreign exchange. The library cooperates with national and academic partner libraries, organisations and institutions from about 50 countries. For exchange purposes the National Library of Armenia purchases all types of publications, mainly books and journals. The list of National Library of Armenia exchanges is available electronically upon request. We are interested in acquiring publications in Armenian, Armenian Studies materials, books and reference materials on humanities, social sciences, business, history, philosophy, philology, linguistics, literature, art and culture, etc.

AUSTRALIA

National Library of Australia
Gift and Exchange, Collection Management Division
A.C.T. Postal Code No. 2600
Canberra

serialorders@nla.gov.au
www.nla.gov.au

The library transmits its own publications and duplicates, but the growth of internet publishing has severely limited the number of publications the library can offer on exchange. The library maintains approximately 1,200 exchange agreements with governmental and non-governmental bodies, including some learned societies and academic institutions, chiefly in the Asia-Pacific region. In general, the library does not actively seek out new exchange arrangements.

The library is a depository library for a number of international organizations, including: Organization for Economic Co-operation and Development, the United Nations Organization and its Specialized Agencies.

The library provides information on exchange of publications generally in Australia but there is no co-ordinating agency for all Australian exchanges.

Publications of the six Australian state governments are not available for exchange purposes from the National Library of Australia. A limited number of Federal government publications are available on exchange to existing exchange partners. These are generally duplicate copies of publications received by the Library.

In return, the National Library selectively seeks materials that fall within the library's collection development policy and are issued by major universities or governmental bodies in the Asia-Pacific region or international organizations.

Directories and other sources: Australian libraries gateway (http://www.nla.gov.au/libraries); National Library of Australia *Collection development policy* (http://www.nla.gov.au/policy/cdp);

AUSTRIA

National Library of Austria

Abteilung für Medienerwerbung und –erfassung/Tausch
Josefsplatz 1, A-1015 Wien
Postfach 308

Tel (+43-1) 53 410 / 414

Fax (+43-1) 53 410 / 445

http:// www.onb.ac.at

The "Internationale Austauschstelle" as it is mentioned in the 4th edition from 1978 doesn't exist anymore - it was closed 1992 and so there is no dispatching centre for other Austrian libraries anymore. The Austrian National Library itself has a lot of exchange contacts like most National Libraries in the world.

BELARUS

National Library of Belarus
9, Chyrvonaarmeyskaja Str.
220636 Minsk

Tel (375 0172) 27 54 63
sol@nacbib.minsk.by
http://kolas.bas-net.by/bla/nb.htm

National Library of Belarus (NLB) provides international exchange of publications

since 1924. The Exchange Section is a part of the Collection Development Department of the library. Three full time librarians are engaged in exchange activities with 340 partners in 51 countries worldwide. The exchange collection consists of monographs, periodicals and other materials printed in Belarus. The library purchases books and other publications for exchange. National Library of Belarus carries out only direct exchanges. Occasionally, it transmits duplicate exchange publications to other Belarussian libraries.

Directory:
Nacyjanalnayja Biblijateka Belarusi: 1922-1992. Bibliographic
Index. Minsk, 1992. 232 p.
Nacyjanalnayja Biblijateka Belarusi. Minsk, 1997. 25 p.
Knizhnoe Dielo. Pressa Belarusi. Minsk, 1999. 736 p.

BELGIUM

Bibliothèque Royale de Belgique / Koninklijke Bibliotheek van België
(The Belgian National Library)
4 blvd de l'Empereur
1000 Brussels

www.kbr.be

BOLIVIA

Archivo y Biblioteca Nacionales de Bolivia
Edificio antiguo: Espanña no. 43
Edificio nuevo: Dalence No. 4
Sucre

Tel (591 4) 645 1481 / 645 2886
Fax (591 4) 646 1208
abnb@mara.scr.entelnet.bo

BOSNIA AND HERZEGOVINA

National and University Library of Bosnia and Herzegovina
Sector of Acquisition, Processing, Periodical and Official Publications
Exchange Department
Zmaja od Bosen St. 8B
Postal code 361
71 000 Sarajevo

The international publications' exchange is one of the most important ways of both supplementing and reconstructing the holdings of the National and University Library of Bosnia and Herzegovina which were destroyed during the recent war. After the six years long break in publication exchange (1991-1997), the Exchange Sector renewed the contacts with cooperating libraries, and along with normalization of the postal services the exchange of publications intensified. Immediately after the war the National and University Library of Bosnia and Herzegovina was exchanging only its own publications since the holdings of duplicates had been totally destroyed. Due to the fact that the library had to reconstruct its own destroyed holdings in extremely inviolable financially and spacewise circumstances, the duplicates' holdings did not exit at all. With huge support from publishers, authors, and colleagues from national and university libraries, Slavic centers, and specially from culture ministries of those countries that recognized our persistence and desire to restitute the destroyed library collection, the National and University Library of Bosnia and Herzegovina managed to restore the function of the State Library.

In the process of restoration of the National and University Library of Bosnia and Herzegovina the international publications' exchange shows the results in the following directions:

- reconstruction of the library's holdings

- promoting Bosnian literature and spreading the national culture outside the borders of Bosnia and Herzegovina

The policy of acquisition foreign literature the library directs to completion of Bosniaca, Slavica, to important works from all scientific fields, to secondary publication from all spheres of science, to scientific and technical systems of information as to the library science literature.

Besides important results achieved through exchanging scientific accomplishments, the National and University Library of Bosnia and Herzegovina would like to show the openness of our country through literature, which mirrors our cultural and scientific development. The cooperation with libraries has been worked out in detail in written agreements on the exchange parameters.

The National and University Library of Bosnia and Herzegovina cooperates with 119 libraries from 39 countries worldwide. The inter-library cooperation is larger with European libraries because it is limited with transportation costs for which our library does not get funded for.

The library keeps the evidence of the exchanged books' value in the balance sheets. The exchange balance has in most times been accomplished through a volume for volume basis or for money value of the publication.

Currently the average yearly exchange in the National and University Library of Bosnia and Herzegovina is approx. 1100 book titles and 60 periodical titles, which satisfies appropriate cooperation, yet does not satisfy the needs of our library.

BRAZIL

Biblioteca Nacional (National Library)

Av. Rio Branco 219-39

22040-008 Rio de Janeiro, RJ

http://www.bn.br/Script/index.asp

BULGARIA

St.St.Cyril and Methodius National Library
Foreign Acquisitions and Exchange Division
88, Vasil Levski blvd.
1037 Sofia

Tel 00359 (2) 988 28 11; 846 60 77
Fax 00359 (2) 843 54 95
kom@nationallibrary.bg
http:// www.nationallibrary.bg

The Foreign Acquisitions and Exchange Division of the National Library assumed the functions of a national exchange centre in 1963 by virtue of a decree and works according to a regulation of the Ministry of Culture dated June 1964. Its activities are financed from the general budget of the National Library. The centre has 1 part-time and 9 full-time employees.

For exchange purposes, the centre disposes of the whole Bulgarian printed production and has financial resources to purchase books and other exchange materials. General exchange materials include publications of universities, academies, learned societies, official bodies as well as duplicates, also slides, maps, prints, microfilms and CDs. It is all direct exchange, mainly for the benefit of the National Library. Transmissions are not undertaken, but information on exchanging partners is given and co-ordination of national exchange exists. The centre works according to the Paris conventions of 1958.

Exchange material received or forwarded is exempt from customs duties.

The National Library maintains international exchange relationships with over 500 libraries and organisations.

CHINA

International Exchange Section
National Library of China
33 Zhong Guan Cun Nan Da Jie
100081 Beijing

CROATIA

Gift and Exchange Division
National and University Library
Ulica Hrvatske bratske zajednice 4
P.O. Box 550
1000 Zagreb

Tel (385 1) 616 41 11
Fax (385 1) 616 41 86
vdolezal@nsk.hr
http://www.nsk.hr

The Gift and Exchange Division was established as a division within the Acquisition Department of the National and University Library. It exchanges publications published by the library, its duplicates but also by purchasing scientific journals and monographs thus meeting our exchange partners' requests, mostly publications dealing with Croatia and its identity, that is, its people and their customs, its language and literature, culture, history and science, in this way fulfilling the primary purpose of exchange.

CUBA

Biblioteca Nacional "José Martí"
Dpto. Desarrollo de Colecciones
Plaza de la Revolución
La Habana
CP 10600 Cuba

Exchanges are made by the exchange department of the National Library which is held in Dpto. Desarrollo de Colecciones.

CZECH REBUBLIC

The Foreign Acquisitions Department
Exchange
National Library in The Czech Republic
Klementinum 190
110 01 Praha 1

www.nkp.cz

DENMARK

The Royal Library (The National Library of Denmark)
P.O. Box 2149
DK-1016 Copenhagen K

http://www.kb.dk

Denmark has ceased any international exchange of publications on the national level. For exchange arrangements with individual libraries, contact the libraries directly.

Directory: *Biblioteksvejviseren*, published by Danmarks Biblioteksforening. Annual. Also electronic version: http://www.bibliotek.dk

DOMINICAN REPUBLIC

Biblioteca Nacional (National Library)
César Nicolás Penson
Santo Domingo

Biblioteca de la Universidad Autónoma de Santo Domingo
Ciudad Universitaria, Apdo 1355
Santo Domingo

ECUADOR

Biblioteca Nacional del Ecuador (National Library)
12 de Octubre 555, Apdo 67
Quito

EGYPT

Egyptian National Library
Sharia Corniche El-Nil
Bulac, Cairo

ESTONIA

National Library of Estonia
Tõnismägi 2
15189 Tallinn

National and parliamentary library.
www.nlib.ee

ETHIOPIA

Addis Ababa University Libraries
P.O. Box 1176
Addis Ababa

FINLAND

Exchange Centre for Scientific Literature
Mariankatu 5
00170 Helsinki

Tel +358-9-22869238
Fax +358-9-22869290
vaihtokeskus@tsv.fi
http://www.tsv.fi

Until 1979 the services of the international Exchange Centre for Scientific Literature were performed by the Library of the Scientific Societies in Helsinki. When it was closed in 1979, the new Exchange Centre for Scientific Literature was founded and forms a division of the Publishing Centre of the Federation of Finnish Learned Societies.

The activities of the Exchange Centre now concentrate on conducting the exchanges of the member societies of the Federation of Finnish Learned Societies. The Exchange Centre is now keeping up some 5400 exchange relations with 3200 exchange partners in 100 countries. On exchange the Centre is sending serial titles and periodicals published by the participating societies, and in return receiving about 7600 serial titles and periodicals, that are forwarded to 120 scientific libraries in Finland.

Exchange service for outside organizations has diminished during last ten years.

Library of Parliament
Acquisition and Collection Service
Aurorankatu 6
FI-00102 Helsinki

Tel. +358 9 4321
kirjasto.kokoelmapalvelu@eduskunta.fi
http://www.eduskunta.fi/kirjasto.

Library of Finnish Parliament was founded in 1872. Since 1913, the services of the Library of Parliament have been to all citizens besides Members of Parliament and parliamentary officials. Today, the library provides information on Parliament, and collects and keeps available a wide range of information on jurisprudence and society, both in print and in digital format. From the very beginning the exchange of official publications with foreign libraries and institutions has been important to the collection development. The Library of Parliament has functioned as an exchange centre of official publications since 1968, after the ratification the Paris convention of 1958, which came in force on 12 May 1968. The exchange is mostly based on mutual agreements between the parties concerned. The exchange material consists mainly of parliamentary documents and official journals.

Directory:

Kirjastokalenteri (annual almanac, 2004 being the current issue, published by BTJ. Kirjastopalvelu)

Gateway to Finnish Research Libraries, http://www.lib.helsinki.fi/tilke/indexeng.html

FRANCE

Bibliothèque Nationale du France
Service des échanges
Quai François-Mauriac
F-75706 Paris Cedex 13

www.bnf.fr

The International Exchange Service, founded in 1877, was attached to the Bibliothèque Nationale in 1936. Its staff consists of twenty full-time employees. It performs exchanges according to numerous direct agreements with foreign national exchange centres and the main libraries all over the world (about 1.500 partners). A list of French publications for exchange purposes in the different matters are sent regularly to them. The centre is also responsible for inter-governmental agreements concerning exchange of official publications entered into by the French government with Germany, Belgium, Canada, Quebec, United States, and United Kingdom.

The International Exchange Service uses for exchange purposes the publications of the Bibliothèque Nationale, duplicates as well as a limited amount of purchase material. Publications are stored for four years.

GEORGIA

National Parliamentary Library of Georgia
Guadiashvili 7
380007 Georgia

GERMANY

Deutsche Bibliothek
Adickesalle 1
D60322 Frankfurt am Main

Tel +49 69 7566204
Fax +49 69 7566476

Staatsbibliothek zu Berlin
Preussischer Kulturbesitz
Potsdamer Str. 33
10785 Berlin

Deutsche Bücherei Leipzig
Deutscher Platz 1
04103 Leipzig
Tel 03 41 / 2 27 10
Fax 03 41 / 2 27 14 44

GHANA

Ghana Library Board
POB 663, Accra

GREECE

The National Library of Greece
Aven Panepistimiou 32
C 10679 Athens

Tel +30 1 3614413
Fax +30 1 3611552

A national exchange centre as such, does not exist in Greece. However a number of Academic Libraries and the National Library of Greece are engaged in such activities. The National Library has signed agreements for exchange of publications with other National Libraries of a number of countries. The National Library of Greece provides also information on exchange opportunities.

GUATEMALA

Biblioteca Nacional de Geatemala (National Library)
5A Avda 7 – 26, Zona 1,
Guatemala City

HUNGARY

National Széchényi Library
Section of Acquisitions
Budavári Palota F. ép.
Budapest, Hungary, 1827

Tel (36 1) 224 3871
Fax (36 1) 202 0804
http://www.oszk.hu

Established in 1951 and reorganized in 1995, the exchange service is a part of the Hungarian National Library and its expenses are included in the budget of this library.

Since 2001 the former International Exchange Service has been incorporated into the Acquisition Department. Despite this change, the exchange activity of the Library has not ended.

In conformity with international library practice, the National Széchényi Library is engaged in a very active book and periodical exchange with foreign collections. Most important exchange partners are the collections of the Carpathian Basin. By this route, the library is able to acquire works which have never entered the market and are thus not commercially available. Equally importantly, this is the usual way in which, the National Széchényi Library acquires works which should clearly belong to the basic collections of the National Library: works published in Hungarian abroad, works published in a foreign language but pertaining to Hungary, and works by Hungarian authors, published abroad in a foreign language. In exchange, the publications of the National Széchényi Library and nationally produced books and periodicals are made available to foreign libraries.

Directory: Könyvtári Minerva 1996, Budapest, 1997, Országos Széchényi Könyvtár, Könyvtártudományi és Módszertani Központ, edited by Ágnes Rácz.

INDIA

National Library
Gift and Exchange Division
Belvedere
KOLKATA – 700027

http://www.nlindia.org/

INDONESIA

Exchange Workgroup, Acquisition Division, Center for Collection Development and Processing of Library Materials and Information Services, National Library of Indonesia
Jl. Salemba Raya 28A
Jakarta.

Tel (62 21) 315 48 63, 315 48 64, 315 48 70
Fax (62 21) 310 35 54
info@pnri.go.id
http://www.pnri.go.id

The workgroup was founded in December 1967. It is a part of Acquisition Division of the National Library of Indonesia and financed by this Department. The staff consists of 5 persons. The Division Exchange Government and Indonesian Publication stresses on the field of social sciences and humanities. From time to time it also purchases other exchange materials. Publications are stored for one year.

The center also operates an exchange services on the basis of bilateral agreement.

IRAQ

National Library
Bab-el-Muaddum
Baghdad

ISRAEL

Jewish National and University Library
P.O. Box 34165
Jerusalem 91341

Fax +02 658 6144
http://jnul.huji.ac.il

The exchange centre is attached to the Hebrew University of Jerusalem. It is managed by the Periodicals Department of the library and funded by the university. The staff consists of one full-time employee.

The centre transmits exchange material to foreign and domestic libraries. It is also responsible for the exchange of official publications and co-ordinates the national exchanges.

According to direct agreements with institutions abroad, it exchanges books and periodicals published by the Hebrew University and by learned societies. Financial resources to purchase exchange material are available.

There are no charges on the importation of exchange shipments and a reduction in the general postal rate is granted.

Directory: *Directory of Special Libraries in Israel*. Tel Aviv, Ministry of Energy and Infrastructure, National Centre of Scientific and Technological Information, 1985. 137 p.

JAMAICA

Circulations Department, Headquarters
Jamaica Library Service
2 Tom Redcarn Drive
Kingston 5
Jamaica, West Indies

Jamaica does not have an exchange centre with national responsibility for exchanges. The following are libraries and institutions which operate formal and informal arrangements:

The Jamaica Library Service does not operate a formal programme for exchanges and gifts. However, several government agencies, associations and benevolent organizations distribute their periodicals gratis to the public through the Jamaica Library Service, as it has a network of service points, throughout the thirteen parishes, islandwide.

The Library of the University of the West Indies, Mona, Kingston 7, Jamaica, operates an exchange and gifts programme (books and periodicals), primarily with other universities, learned societies and the United States Library of Congress. The Acquisitions Department is in charge of this programme. One para-professional and four non-professionals assist with the programme. The material exchanged consists mainly of West Indian government publications (excluding Jamaica) and university publications. Funds are provided from the general book vote at the discretion of the University Librarian.

The National Library of Jamaica does not have a formal programme. If the Library receives extra copies of publications, these are donated to relevant libraries.

The Planning Institute of Jamaica (PIOJ), which is situated at 10 Grenada Way, New Kingston, Kingston 5, operates an exchange and gifts programme. Locally, they exchange with the Statistical Institute of Jamaica (STATIN), the Ministry of Foreign Affairs, United Nations Development Programme (UNDP) and all the libraries belonging to the Social & Economic Information Network (SECIN). Also, a copy of each publication is deposited with the National Library, Jamaica's national depository for literary works. Donations are also made locally to the Media Houses, Government Ministries, the Cabinet and the Permanent Secretaries.

Internationally, exchanges are made with the following: The World Bank, International Development Bank (IDB) and the World Trade Organization (WTO). Donations are made to the Prime Minister of Barbados. Two librarians are in charge of this programme and funds are provided from the operating budget. The material exchanged and donated are socio-economic in content. Some of these publications are accessible online, gratis.

The Jamaica Printing Services, situated at 77 1/2 Duke Street, Kingston, is neither a government institution nor an agency anymore, although it prints material for the government. The Jamaica Printing Services donates material *on request* for the following categories of institutions: voluntary organizations, schools, churches and communities. However, it does not have an exchange programme.

The Statistical Institute of Jamaica (STATIN), situated at 9 Swallowfield Road, Kingston 5, operates an exchange programme. Locally, the Institute exchanges statistical publications with the Bank of Jamaica (BOJ), the Ministry of Labour, the Planning Institute of Jamaica (PIOJ), the Ministry of Finance and the Ministry of Health. It also deposits a copy of each publication with the National Library of Jamaica.

Otherwise, STATIN exchanges statistical publications with various statistical agencies within the Caribbean and internationally, such as the United Nations and the International Monetary Fund (IMF). A clerk handles this programme and funding is obtained through the operating budget.

The Bank of Jamaica does not have a formal programme. However, all the Bank's publications are given gratis to all institutions who request them, *nationally* and *internationally.* Moreover, these publications are available gratis on the INTERNET. The address of its website is: www.boj.org.irn Some of the institutions who request publications from the Bank are: Central Banks - worldwide, the IMF, the World Bank and Universities - both locally and internationally. An administrative assistant, a typist and a clerk are in-charge of this programme and funds are provided from the operating budget. The publications are financial and economic in content.

The Supreme Court does not have an exchange and gifts programme in force anymore.

The Libraries of the University of Technology, the National Caribbean University and all special libraries, with the exception of those listed previously, operate informal arrangements.

JAPAN

Foreign Material Acquisition Division, Acquisition Department
The National Diet Library
1-10-1 Nagata-cho
Chiyoda-ku
Tokyo 100-8924

Tel (81 3) 3581 2331
http://www.ndl.go.jp

The National Diet Library (NDL) has been functioning as an international exchange center of publications since its establishment in 1948 when it took over the Japanese government's international exchange project that had been conducted by the Ministry of Foreign Affairs since 1875. The National Diet Library was officially designated for the "national exchange service" when Japan ratified two UNESCO conventions in 1984.

The National Diet Library has exchange relations with most countries in the world, currently conducting exchanges with over 900 institutions including the international organizations.

The National Diet Library mainly deals with official publications. The international exchange of official publications falls into three categories, namely: "blanket exchange", "selective exchange" and "specified exchange".

"Blanket exchange" is a system whereby major official publications are comprehensively exchanged based on agreements between governments or conducting institutions. Its present partners are: Library of Congress (USA), National Library of Australia, Berlin State Library - Prussian Cultural Heritage (Germany), National Library of Canada (shipped to University of British Columbia Library), University of California at Berkeley (USA), National Central Library (Taiwan).

"Specified exchange" means to exchange basic official publications mutually specified according to an agreement between governments or the institutions concerned. This exchange is more limited in scale than the "blanket exchange".

For exchange with many institutions, the method called "selective exchange" is used, that is to exchange materials selected by each other with consent.

Besides official publications, the National Diet Library purchases commercial publications wanted by its partners and exchanges them based on the principle of exchange in equal amount or equal price. In particular, exchange of national literature is important. That is, the library acquires Japonica, materials on Japan published in other countries, while to those countries it sends materials on them published in Japan.

The National Diet Library is designated as the depository library by several international organizations and receives their publications. The National Diet Library is now designated by: UN, ILO, UNESCO, ICJ, IMF, WHO, FAO, ICAO, IAEA, OECD, World Bank, Asian Development Bank (ADB), International Maritime Organization (IMO), and others.

Source: *Kokuritsu Kokkai Toshokan Nenpo* (Annual Report of the National Diet Library 2001)

JORDAN

National Library of Jordan
P.O.BOX 6070
Amman 11118

Tel (962 6) 4610 311
Fax (962 6) 4616 832
nl@nic.net.jo
http://www.nl.gov.jo

Besides exchange section located in the Library of the University of Jordan, the department of the National Library has a section that handles the responsibility of exchange and gifting. It exchanges publications with nearly 114 national and international libraries (schools, universities, official establishments and other organizations). The section receives nearly 2000 publications and presents 7500 in exchange yearly.

KAZAKHSTAN

The National Library of the Republic of Kazakhstan
Abai av., 14,
480013 Almaty

Tel (3272) 69 65 86
Fax (3272) 69 65 86
http://www.nlrk.kz

The National Library of the Republic of Kazakhstan (NLRK) is one of the largest libraries of the CIS countries under the authority of the Ministry of Culture,

Information and Public Concord of the Republic of Kazakhstan. It was founded in 1910. Today it is the acknowledged leader in the librarianship of Kazakhstan. In the field of librarianship the National Library is the leading scientific and research, bibliographic reference, information and library automation centre. Since 1931 the National Library receives legal deposit, published in Kazakhstan and functions as the state national book store in the republic, which acquires and holds documentary written heritage of the Kazakhs and other peoples living in Kazakhstan.

The National Library pays great attention to the development of new and supports traditional cultural interlinks with foreign countries.

For many years the National Library keeps exchange with 200 libraries of the CIS countries and 80 cultural organizations from 40 countries of the world. Among them are such libraries as the Library of Congress, Harvard University Library, libraries of India, Spain, Egypt, France, Great Britain, Norway, Austria, Germany, Iran, Turkey, China, and Korea.

In 1992 the National Library became a member of IFLA (International Federation of LibraryAssociations) and took an opportunity to take part in IFLA professional programs.

KOREA, SOUTH

Support and Cooperation Division
The National Library of Korea
San 60-1 Banpo-dong
Secho-gu
Seoul 137-702

http://www.nl.go.kr

The Support and Cooperation Division responsible for exchanging publications abroad is attached to the National Library of Korea. The Library is a legal deposit center for domestic books and builds the National Bibliography records.

One of the main functions of the Division is to exchange government publications and general books with 262 libraries, institutions, etc. abroad. The staff consists of 12 persons employed full-time and part-time (when necessary).

Exchange shipments are exempt from custom duties and postal reductions are granted for printed matter. The library has focused its effort on extending the contents of digital materials in full text. The number of full text databases created by the library is 11, in which 275.000 titles where digitalized. They are available at the library's Web site.

LATVIA

LatvianAcademic Library
R½pniecÓbas iela 10
1235 Riga

Latvian National Library
Kr. Barona iela 14
1423 Riga

LIBYA

National Library of Libya
POB 9127
Benghazi

LITHUANIA

Martynas Mazvydas National Library of Lithuania
Acquisition of Foreign Documents Division
Gedimino pr.51
2635 Vilnius

Tel +370 2 629023
Fax +370 2 627129
biblio@lnb.lrs.lt
http://www.lnb.lt

Till 1954 international exchanges were centralized in the Soviet Union. Correspondence and printed matters with exchange partners were turned to the main libraries of Moscow and Leningrad and they used to send these printed matters to according

foreign libraries.

Since 1954 the government of the Soviet Union granted a right for some libraries of allied republics to be on publications' exchange with foreign libraries directly. The right to be on directly exchange was granted only to the Library ofAcademy of Sciences of Lithuania, to the Republic (now the National) Library of Lithuania and to the Library of University of Vilnius.

Up to now these three libraries are the leaders of international exchanges in Lithuania. At present the situation approximately is a such: the Library of Academy of Sciences of Lithuania has about 600 exchange's partners in 41 countries of the world. Mainly there are publications from the sphere of precise and fundamental sciences, also with publications of 15 research institutes of Academy of Sciences.

Martynas Ma_vydas National Library of Lithuania has about 200 exchange's partners in 36 countries of the world. Mostly its exchange's partners are National and Parliamentary libraries which are on exchange for the most part of humanitarian, social, politics and law sciences literature.

The library of the University of Vilnius has about 300 exchange's partners in 40 countries of the world. Its partners mainly are Universities and the other scientific institutions. They are on exchange with literature of universal profile.

Financing. Libraries purchase the literature for international exchange from the means divided for acquisitions of the library in general.

Executives. There are groups of international exchange of 3 - 4 persons at the acquisition department.

Foreign partners of these three above mentioned libraries are the largest libraries of America and Europe - near 20 - 30%. The sent and received literature almost never repeats, we often associate and consult between ourselves.

However, we haven't a Centre of International Exchange in Lithuania yet.

LUXEMBOURG

Bibliothèque Nationale (National Library)
37 blvd F. D. Roosevelt
2450 Luxembourg

MACEDONIA

The National and University Library "St Kliment Ohridski"
Bul. Goce Delcev, 6 PO Box 566
1000 Skopje

Tel (389 91) 115 177, 133 418
Fax (389 91) 226 846
exchange@nubsk.edu.mk
http://www.nubsk.edu.mk

Sector for Accessing and Processing of Library Materials, Acquisition Department

The section exchanges the publications of the National and University Library of Macedonia, duplicate holdings and publications purchased for that purpose, and it sends abroad copies of the publications printed in Macedonia. In 2002 it received 1310 exchange books, dispatched 4618 and

exchanged 213 periodical titles. The service for exchange of publications operates within the Acquisition Department, carries out the exchange of monographic and periodical publications. Today the library cooperates with 255 national and university libraries and other institutions from 73 countries. An easier and faster communication and co-operation is available because of the use of Internet. The exchange list with some libraries are being sent by e-mail. The section also provides information concerning exchange possibilities. It carries out bibliographical activities by publishing bulletins listing new acquisitions.

MALAWI

The National Archives of Malawi
P.O. Box 62
Zomba

Tel (265 01) 525 240
archives@sdnp.org.mw

Although at the moment there is no one centre which acts as a national exchange centre, the National Archives of Malawi enjoys exchange programmes with all libraries within Malawi and with 41 institutions worldwide. The National Archives is the publisher of the Malawi National Bibliography and acquires copies of items published by various government departments through the Printed Publications Act. This arrangement qualifies the National Archives to be a depository library and a custodian of all government publications and have surplus material, which are used as exchange publications together with the Malawi National Bibliography.

The Government publications are also distributed to interested parties as gifts and donations. Due to escalating postage charges the National Archives may ask for a token contribution from interested exchange partners to cover postage. No custom duties are levied for exchange shipments.

MALAYSIA

The National Library of Malaysia
Malaysiana Information Center
PDS Coordinating Center
232, Jalan Tun Razak
50572 Kuala Lumpur

brm@pnm.my
htpp://www.pnm.my/new

The National Library of Malaysia is a Federal Government Department, established in 1972. Among others, it acts as the National Coordinating Centre for the Document Delivery System whereby, a unit was set up in 1988 to coordinate national and international lending and exchange of library materials among member libraries and/or institutions. It acts as the centres for the transmission of publications received from various national and international organizations, learned societies, resource centres and libraries.

The Centre handles all shipments to foreign countries and consignments dispatched and received are exempted from customs duties and sales tax. A postal reduction is granted for printed items.

The National Library of Malaysia maintains numerous bilateral exchange agreements with universities, research and government institutions and exchanges research and official publication as well as duplicates and surplus materials.

MALTA

The National Library of Malta
36, Old Treasury Street
Valletta CMR 02

Tel +356 22 43 38
Fax +356 23 59 92

Although at present, no national exchange centre exists, the Acquisition Section within the Bibliographic Services Unit conducts exchanges with other national libraries in various countries. For the most part, national bibliographies are exchanged. However, from time to time, the National Library of Malta receives donations of publications of historical and cultural interest from the Ministero per i Beni e le Attività Culturali, Italy.

No postal reductions is granted, no taxes are levied.

MOLDOVA

The National Library of the Republic of Moldova
78 A, Str. 31 August 1989
MD-2012 Chisinau

Tel +373 2 22 1475; 24 1096; 24 0630

The section of the International Exchange is a part of the National Library of Moldova. It has a staff of two full-time employees.

The Exchange material consists of monographs and periodicals published in Moldavian, Romanian, Russian, English and French languages. The library can purchase books and other materials for exchange. It carries out only direct exchanges. The main interest of the library exchange lays in humanities, social sciences and literature .

MOROCCO

Bibliothèque Générale et Archives
PB 1003, Ave Ibn Batouta
Rabat

NETHERLANDS

Koninklijke Bibliotheek (Royal Library)
POB 90407
Prins Willem-Alexanderhof 5
2509 LK The Hague

NIGERIA

National Library of Nigeria
4 Wesley St.
PMB 12626, Lagos

NORWAY

The National Library of Norway
Oslo Division
Exchange Office
P.O. Box 2674 Solli
N-0203 OSLO

Tel (47) 2327 6028
Fax (47) 2327 6010
exchangeoffice@nb.no
http://www.nb.no

The Exchange Office at the National Library of Norway is exchanging publications with libraries all over the world on behalf of the National Library of Norway and the University of Oslo Library, Library of Arts and Social Sciences. It was originally the exchange office of the University of Oslo Library, and was established as far back as in 1845. The office is partly financed by the National Library of Norway and partly by the University of Oslo Library.

The Exchange Office forwards publications issued by official and governmental agencies, universities, learned societies, and the National Library of Norway. Lists of duplicates are sent to some of the exchange partners 6-10 times a year. The office also has a small amount at its disposal for purchasing exchange materials. The office is engaged in direct exchanges according to agreements with similar institutions; it has agreements with other national libraries to receive Norvegica extranea in exchange for similar material published in Norway, and intergovernmental agreements with countries like U.S.A. and Germany.

Before 2000 the Exchange Office effected the distribution of doctoral dissertations from the University of Oslo, and it has still a stock of dissertations up to the year 2000 available for the exchange partners.

Two persons are working in the Exchange Office, one full time, and the other part time.

Directory: *Norske vitenskapelige og faglige biblioteker : en handbok.* Utgitt av Riksbibliotektjenesten. 9. utg. Oslo : RBT, 1999.

PANAMA

Biblioteca Nacional (National Library)
Apdo 7906
Panama City 9

www.binal.ac.pa

PHILIPPINES

Collection Development Division (CDD), Exchange Section
The National Library of the Philippines (TNLP)
T.M. Kalaw Street, Ermita 1000
Manila

The Collection Development Division (CDD), Exchange Section of the National Library of the Philippines acts as the exchange centre for official and other Philippine Publications. The Library obliges all official bodies to supply up to 200 (or more) copies of all documents and publications, printed, issued or ordered by them to The National Library for exchange purposes. The division distributes among local beneficiaries the materials received from the United States and other exchange institutions.

The Collection Development Division Exchange Section employees work on the exchange activities, sends official publications and other Filipiniana references to foreign and local exchange partners and receives foreign publications, acknowledges, processes and distributes them to the research areas for reference purposes.

The Library offers for exchange official publications (Philippine National Bibliography PNB both in book and CD ROM format) and other Filipiniana materials only, but receives from foreign exchange partners all types of publications in English language or other languages with English translation only. Exchanges are carried out under formal or informal agreements with both foreign and local institutions, colleges and universities.

Electronic publishing makes exchange activities easier particularly in shipping as well as in using and maintaining the materials.

POLAND

Biblioteka Narodowa

The International Exchange Bureau for Publications

Al. Niepodlegzoúci 213

02-086 Warszawa

Tel (48 22) 608 2569

http://www.bn.org.pl

bmww@bn.org.pl

The International Exchange Bureau for Publications, incorporated by the National Library from the Ministry for Foreign Affairs on August the 1st, 1931, is a department of the National Library since 1928 functioning according to the decisions of the Paris Convention from 1958.

To the competency of the Bureau belong:

1. Exchange of addressed publications (sending publications from Poland, obtaining foreign publications and transmitting them to domestic institutions).

2. Exchange of unaddressed publications (on basis of agreements with other domestic and foreign libraries and institutions).

3. Support and facilitation of exchange of official and academic publications between Polish and foreign libraries and institutions.

4. Giving information on international exchange of publications.

5. Realizing information retrievals and orders of Polish and foreign libraries concerning international exchange of publications and legal documents, observance of which are obliged departments of international exchange publications of libraries and centres of information of research institutes.

The main task of the International Exchange Bureau for Publications is to free (without cost) exchange official and academic publications between Polish and foreign libraries and institutions on the basis of the Paris Convention, which

Poland joined in 1970. The fact of joining the Paris Convention was documented and published in *Dziennik Ustaw* nr 8 of April the 13th and in the Appendix of this issue. In the light of its regulations, the central institution responsible for the realization of international exchange was the International Exchange Bureau for Publications. In the beginning it was situated in the Library of the Ministry for Foreign Affairs, then for a short time in the building of the Sejm, until eventually the authorities decided that an appropriate place for it would be the Ministry of Culture. In various discussions and debates it was decided, that the most appropriate place would be the National Library, so the Bureau was joined to the Department for Collections Development of the of the National Library. In 1989, after the plan for placing the Bureau in the new building was confirmed, it was given a permanent place in the basement. In the beginning, the staff consisted of seven employees, but with time it was reduced.

The year 2000 was a break-through year for the Bureau — because of organizational changes. Traditional working methods were abandoned and the work was completely automated. There emerged the International Exchange Bureau for Publications Database supported by the system MAK and including information on: titles, addresses of publishers, distributers and vendors, issues, document types and other information facilitating registration.

The International Exchange Bureau for Publications receives materials almost from the whole world: from national libraries, university libraries, learned societies and other publishing institutions. The tasks of the Bureau are the transmission of material to appropriate and permanent partners in Poland, mainly to central public offices and libraries, and the acquisition of new material. Foreign partners obtain on exchange mainly official publications, such as parliamentary publications, but not entirely. Every now and then the Bureau receives publications, which are sent to appropriate libraries as gifts after been checked in the catalogue of the National Library. Receiving material, registration and later dispatch of it are of great importance to regular functioning of the Bureau.

The International Exchange Bureau for Publications is trying to find new partners for co-operation and in this way to realize the decisions of the Paris Convention. The Bureau co-operates with 95 exchange centres in Poland and abroad.

ROMANIA

RomanianAcademy Library
International Exchange Department
Calea Victoriei 125
010071 Bucuresti

Tel (401) 314 2434
Fax (401) 312 3381
go@bibnat.ro
http://www.bibnat.ro

The first exchange relationships were established 1870, when the RomanianAcademic Society received a few collections of periodical publications from the Danish Academy of Sciences in exchange for 26 volumes published by the Society. In 1874, at writer Alexandru Odobescu's proposal, a systematic exchange program with foreign scientific institutions was initiated. The first exchanges with institutes from the United States date back to1883. Until 1952 the Romanian Academy Library was the National Centre for Exchange of Publications of Romania.

The exchange program expanded during the time so that, while at the end of World War I the Romanian Academy Library had 145 exchange partners, there were 850 active partners all over the world at the beginning of the World War II. Presently, the academic exchange system includes approximately 15 000 partners from all scientific branches. In 1949 a separate Exchange Department was founded, and it took over the tasks of other departments and is currently handling the exchange program.

In exchange for the foreign publications the library offers the Romanian Academy publications, edited by the Romanian Academy Publishing House.

The Exchange Department is lead by the Scientific Council of the Library and financed by the Romanian Academy. It consists of 10 specialized full time persons.

Among exchange partners there are similar academic institutions, university, scientific and public libraries according to the agreements concluded, many of them, even from the beginnings of the exchange program. Beyond sending the exchange publications, the department supplies information about exchange possibilities and coordinates the exchanges of the Academic Institutions' Reviews. The exchange department has also a librarian, who handles the international loan of publications.

RUSSIA

National Library of Russia
18 Sadovaya Street
191069 St. Petersburg

Tel 00 7 812 110 6253
Fax 00 7 812 310 6148
rnb@glas.apc.org

Russian State Library
Vozdvigenka 3
101000 Moscow

Tel +7 095 202 35 65
Fax +7 095 200 22 55
main@irgb.msk.su

RWANDA

Bibliothèque de l'Université Nationale du Rwanda
B.P. 117 Butare

The centre was established in 1964 as a part of the National University of Rwanda. It is financed by the Rwandese Government. It has a full-time staff of professional librarians.
The centre exchanges books and periodicals originating from the university and, occasionally, some duplicates and surplus material. It does not exchange official publications. Apart from direct exchanges, it transmits exchange material to other libraries.

SAUDI ARABIA

King Saud University Libraries
POB 22480
Riyadh 11495

SENEGAL

Bibliothèque Centrale
Université Cheikh Anta Diop de Dakar
BP 2006 Dakar

Tel (221) 824 69 81
Fax (221) 824 23 79
http://www.bu.ucad.sn

The library (founded 1952) is fully automated and has access to Internet. The library has exchanges with university libraries and national libraries in Africa and other parts of the world, mainly in Europe and North America. The materials exchanged consist mainly of academic and university publications (thesis and articles).

The exchanges of the library are made through the Reference Service of the Library and we use the international regulations for interlending and document supply. The Library is the most important in Senegal and play the role of national exchange centre in Senegal.

SINGAPORE

National Library
Gifts and Exchanges
Library Supply Centre
No. 3, Changi South St. 2
Tower B, Level 3
Singapore 486548

Fax: 6546-7262
E-mail: gifts_exchanges@nlb.gov.sg
www.lib.gov.sg/

SLOVAKIA

Matica slovenská (Slovak National Library)
L. Novomenského 32
03652 Martin

http://www.matica.sk/

SLOVENIA

National and University Library
Acquisition Department
Turjaska 1
P.O. Box 259
SI - 1000 Ljubljana

Tel (01) 2001 110
Fax (01) 4257 293
http://www.nuk-uni-lj.si

Since the independence of Slovenia, the National and University Library in Ljubljana (NUL), has continued with the exchange relationships which were established in the former Yogoslavia. The programme of the presence of the Slovene publications abroad is financed by the Ministry of Culture of Slovenia. NUL is appointed to provide the most important Slovene publication and information on the Slovene press.

The exchange is handled by the Acquisition Department. There is one exchange librarian employed full time and an acquisition librarian helps in correspondence with the partners. NUL exchanges its own publications, duplicate holdings and publications purchased for that purpose. These are mainly the publications in the social sciences and humanities, literary works of acknowledged authors (especially the editions including the studies on authors) and official publications. The Slovene dissertations are not available on the exchange basis as only few copies of one title are prepared by the author. NUL publishes the Slovene National Bibliography and this title is dispatched to the majority of the current partners (135 libraries).

The exchange material is sent to the national and university libraries. NUL cooperates with more than 150 libraries in all continents, but mainly with the libraries in Europe and North America. In the recent years, some foreign libraries have ended the exchange agreements due to lack of funds, change of collection development policy or acquisition of material by purchase rather than by exchange. However, the majority of exchanges are still carried on. Duplicates are offered by means of lists arranged by subjects. NUL also offers irregularly lists of recommended "Slovenica" arranged by subject. 1100 titles of serial publications and 1200 books are supplied annually to the foreign libraries. Daily newspapers are supplied directly by the publishers on the daily basis.

In return, NUL receives foreign publications in the social sciences and humanities, official publications and the Slovene publications published abroad (the Slovene author, publications referring to Slovenia or the Slovenes).

The exchange librarian keeps a file card of the shipments and since 2002 a new automated module for exchanges has been introduced. The balance of material sent and received is based mainly on the monetary value but with some libraries other methods are used as well. The postage costs are covered by NUL.

SPAIN

Biblioteca Nacional
Servicio de Compra, Canje y Donativo
Sección de Canje
Paseo de Recoletos, 20-22
28071 Madrid

Tel (34 1) 580 78 00
Fax (34 1) 577 56 34
http://www.bne.es

Established in 1930, the former Exchange Service was originally attached to the Servicio Nacional de Lectura. Since 1957, it has worked as an independent organization in close cooperation with the Exchange Service of the National Library and with the Publications Department of the Junta Técnica de Archivos y Bibliotecas. On 1987 it was attached to the National Library as a Section in the Service of Purchase, Exchange and Donation and into the Acquisition Department.

The Exchange Section works according to the Unesco Conventions on the exchange of publications of 1958. The staff consists of three persons employed full time. There are exchange connections with 410 libraries, mainly in Europe and America.

Besides direct exchanges, the section forwards exchange material from official and scientific institutions in Spain to the respective centres abroad. It provides information concerning exchange opportunities. The section exchanges books and periodicals, using for this surplus material of the National Library and its own publications. On the other hand, the publications obtained in the same way, are added to the collections of the National Library.

Nowadays the work routines are changing and it is possible to connect to other libraries via e-mail and also to watch the exchange lists from the different institutions via Internet, which is, of course, better and faster.

Postal reductions for books are granted.

SRI LANKA

National Library and Documentation Services Abroad (NLDSB) and National Library and Documentation Centre (NLDC)

No 14, Independence Avenue

Colombo 7

nldsb@mail.natlib.lk

nldc@mail.natlib.lk

http://www.natlib.lk

The National Library of Sri Lanka was established in April 1990 and was renamed as the National Library and Documentation Centre (NLDC) under act no. 51 of 1998. The Board is responsible to the Ministry of Education. One of the functions of the National Library is promoting and maintaining national and international library cooperation. NLDC is the national exchange centre in Sri Lanka. International Exchange Program functions under the Acquisition Division of NLDC. At present the International Exchange Program has been established with 56 foreign national and special libraries, universities, and similar institutions. The Sri Lanka National Bibliography (monthly) and the Library News (quarterly) and Natnet Lanka Newsletter (biannually) are distributed among the International Exchange Program participants.

Titles of interest are selected from a list of publications sent to us periodically by these institutions. We select titles specially on Sri Lanka, art, culture, library and information science, science and technology, etc. Books, periodical, atlases, micro forms, CD-roms and computer diskettes are received under this program. In similar manner lists are sent by us periodically to these institutions for selection. Annually NLDSB allocates Rs. 50, 000,- (approx.) for this program. A local exchange program also has been established with a selective number of institutional and academic libraries in Sri Lanka.

SWITZERLAND

Switzerland does not have a National exchange centre. The National Library does not make any exchanges.

TAIWAN

National Central Library
Bureau of International Exchange of
Publications
20, Chungshan S. Road
TAIPEI 10040

www.ncl.edu.tw/english/english.htm

THAILAND

The National Library of Thailand
Samsen Road
Bangkok 10300

Tel. (02) 628 5180

The Centre is attached to the Department of Fine Arts of the Ministry of Culture. It is managed and financed by the National Library of Thailand. It carries out the exchange of duplicates and publications of national official bodies.

The main functions of the Centre are the maintenance of direct exchange and co-ordinations of the national exchanges.

All incoming and outgoing exchange consignments are exempt from customs duties. No reduction of tariffs is granted.

The Centre executes several inter-governmental exchange agreements (with the United States, the United Kingdom and Australia) and with the United Nations Agencies. Up to 2003, the total of international exchange agencies is 75 nations, 242 institutes.

TUNISIA

Bibliothèque Nationale de Tunisie
Service des Acquisitions, Bureau des Échanges et Dons
20, Souk el Attarine
B.P. 42
1000 Tunis

TURKEY

Millî Kütüphane
Dis Iliskiler Subesi
National Library of Turkey
Exchange and Loan Service
Bahçelievler Son Durak
06490 Ankara

Tel (90 312) 212 6200
Fax (90 312) 223 0451
disilis@mkutup.gov.tr
http://web.mkutup.gov.tr

Director of the International Exchange and Loan Service: Miss Belgin Karahan,

Exchange and Loan Service was launched in 1949, as a small department within the body of the National Library, established in 1946. It is financed by the government within the budget of the National Library. Its staff consists of six full time employees managed by the section manager.

The department does not act as intermediary between the libraries in Turkey and/ or abroad, but performs direct exchanges. It is not responsible for the coordination of the national exchanges, and does not supply information about exchange opportunities.

Within the frame of cooperation between libraries, the Service provides the users in Turkey and abroad access to the documents and information which they need, and in this respect carries on publication exchange with 122 libraries abroad. Similarly, in the frame of Cultural Exchange Programs among the national libraries and universities of 80 countries, the Service carries out the loaning processes of books, periodicals and their microfilms and photocopies. It sends 251 libraries and universities the Türkiye Bibliyografyasi (Turkish National Bibliography) and

Türkiye Makaleler Bibliyografyasi (Bibliography of Articles in Turkish Periodicals) which are produced by the National Library of Turkey.

UKRAINE

Vernadsky National Library of Ukraine
Pr. 40-richya Zhovtnya 3
252039 Kiev

National Parliamentary Library of Ukraine
Holosijevski pr. 84
03040 Kiev

UNITED KINGDOM

The British Library no longer operates a national exchange centre.

UNITED STATES OF AMERICA

Library of Congress Exchange Programs

The Library of Congress maintains exchange arrangements with approximately 5.000 institutions throughout the world. The majority of these arrangements are handled as part of the Library's Duplicate Materials Exchange Program. In addition, a large number of other exchanges are administered through the Library's overseas field offices. The Library of Congress also maintains less than a hundred official exchanges through the International Exchange Service, managed by the Government Printing Office, in which full or partial sets of United States Government documents are sent to foreign governments and institutions.

Library of Congress exchanges are established by the divisions of the Acquisitions Directorate, which are based on geographic areas of responsibility. If an institution is interested in establishing an exchange arrangement with the Library, it should contact the appropriate divisions chief as show below.

From the United States, Australia, Canada, Ireland, New Zealand, Oceania, Papua-New Guinea, United Kingdom, contact:

Anglo/American Acquisitions Division (LS/ACQ/ANAD)
Library of Congress
101 Independence Avenue, S.E.
Washington, D.C. 20540-4170

Tel 202-707-5361
Fax 202-707-9440

From the countries of Africa and Asia and to inquire about the Cooperative Acquisitions Programs, contact:

African and Asian Acquisition and Overseas Operations Division (LS/ACQ/AFAOVOP)
Library of Congress
101 Independence Avenue, S.E.
Washington, D.C. 20540-4150

Tel 202-707-5273
Fax 202-707-4445

From the countries of Europe, Latin America, and the Caribbean, contact:

European and Latin American Acquisition Division (LS/ACQ/ELAD)
Library of Congress
101 Independence Avenue, S.E.
Washington, D.C. 20540-4180

Tel 202-707-5243
Fax 202-707-2086

UZBEKISTAN

National Library of Uzbekistan
Horezmskaya 51
Tashkent 700047

navoi@physic.uzsci.net

The National Library of Uzbekistan directs the International Exchange Department since 1957. The International Exchange Department Sector of Foreign Literature Department performs the duties of the International Exchange Department. The International Exchange Department is the main source of acquisition of the National Library with foreign literature. Subjects of such literature correspond to the library's type. The National Library is the leading library on acquisition of humanitarian type literature. Because of the rise in price of printing production and postage in 1993, sending of literature to foreign partners has been terminated temporarily.

At present, the library has exchange connections with 15 organizations from 7 countries of the world. The department has available exchange materials for the International Exchange Department's purposes, which consists of periodicals and monographs, published in Uzbekistan in Uzbek and Russian.

VATICAN CITY

Vatican library
Acquisition & Exchanges
Cortile del Belvedere
00120 Città del Vaticano

Tel (39 06) 6987 9402
Fax (39 06) 6988 4795
bav@vatlib.it
http://bav.vatican.va

SELECTED BIBLIOGRAPHY

Allardyce, Alex: *Letters for the international exchange of publications : a guide to their composition in English, French, German, Russian and Spanish.* Munchen : K.G. Saur, 1978. (IFLA Publications ; 13).

Barker, Joseph W: Gifts and exchange. In: *Technical services today and tomorrow.* Ed. Michael Gorman. Englewood, Colo. : Libraries Unlimited, 1990. Pp. 23-37. (2nd ed. 1998.)

Buckland, Michael: *Information and information systems.* New York : Praeger, 1991.

Clark, Mae: Gifts and exchanges. In: Schmidt, Karen: *Understanding the business of library acquisitions.* Chicago : American Library Association, 1990. Pp. 167-185.

Cline, Hugh F. and Loraine T. Sinnott: *Building library collections : policies and practice in academic libraries.* Lexington, Mass. : Lexington books, 1981.

Collection management for the 21st century : a handbook for librarians. Ed. by G. E. Gorman and Ruth H. Miller. Westport, Connecticut : Greenwood Press, 1997.

Conférence sur les échanges internationaux de publications en Europe. Budapest, 13-19 septembre 1960. Budapest, 1962.

Deal, Carl: The administration of international exchanges in academic libraries : a survey. In: *Libraries acquisitions : practice & theory* 13 (3), 1989.

The gifts and exchange function in ARL libraries : a SPEC kit. Compiled by Catherine Denning. Washington, DC : Association of Research Libraries, Office of Leadership and Management Studies, [1999].

Handbook on the international exchange of publications. 2nd ed. Ed. Gisela von Busse. Paris, 1956.

Handbook on the international exchange of publications. 3rd ed. Ed. Gisela von Busse. Paris, 1964.

Handbook on the international exchange of publications. 4th ed. Ed. Frans Vanwijngaerden. Paris, 1978.

Hogg, Ron: The death of exchange. In: *Slavic and East European information studies*, Vol. 3, No 2/3 2002. Pp. 29-43.

Hogg, Ron: Exchange of publications or electronic databases?. In: *10th Anniversary International Conference 'Crimea 2003' Libraries and Associations in the Transient World: New Technologies and New Forms of Cooperation. Conference Proceedings*, Vol. 2. Pp. 485-489. Moscow : GPNTB, 2003.

The international exchange of publications : Proceedings of the European conference held in Vienna from 24-29 April 1972. Ed. Maria J. Schiltman. Pullach/ München, 1973.

Lane, Alfred H.: *Gifts and exchange manual.* Westport, Conn. : Greenwood, 1980.

Manuel des échanges internationaux de publications. Préparé avec le concours de Mlle J. L. Dargent. Paris, 1950.

Metz, Johannes: International exchange of official publications. *INSPEL* 34 (2000) 2, pp. 80-89 and http://www.ifla.org/IV/ifla65/054-83_e.htm

Shaw, Margaret J: Dinosaur or dynamic: exchange of art publications in the digital age / 65th IFLA Council and General Conference, Bangkok, Thailand, August 20 - August 28, 1999. http://www.ifla.org/IV/ifla65/papers/051-107e.htm

Studies in the international exhange of publications. Ed. by Peter Genzel. München, 1981. (IFLA Publications, 18).

Unesco handbook of international exchanges. 1-2. Unesco, 1965 and 1967.

http://www.ala.org/alcts/organization/div/deu/brochure.html

Discussion list: GIFTEX-L. To order: listserv@lsv.uky.ed - subscribe giftex-l [first name.surname]